T0345277

Ground Improvement for Coastal Engineering

This practical guide covers the investigation, design, and execution of ground improvement in coastal areas. It explains how to decide whether ground improvement is necessary, which method to choose, and how to design and execute it.

Recognising the soft ground commonly found in coastal areas, the book introduces various ground improvement technologies including seismic reinforcement and liquefaction countermeasures and addresses the measures to be taken to sustain ground against external forces. Reliable Japanese ground improvement technologies are presented as well as the latest Building Information Modelling (BIM)/Information and Communication Technology (ICT) technology used in their execution. The book also includes measures that can be taken against contaminated soil and considers ground improvement design on site.

- Unique focus on coastal applications
- Summarises leading-edge Japanese practice

The book suits professionals in the ground improvement industry, especially geotechnical designers and contractors.

Ground Improvement for Coastal Engineering

Hidenori Takahashi

CRC Press
Taylor & Francis Group
Boca Raton   London   New York

CRC Press is an imprint of the
Taylor & Francis Group, an **informa** business

Cover image: Hidenori Takahashi

First edition published 2023
by CRC Press
6000 Broken Sound Parkway NW, Suite 300, Boca Raton, FL 33487-2742

and by CRC Press
4 Park Square, Milton Park, Abingdon, Oxon, OX14 4RN

CRC Press is an imprint of Taylor & Francis Group, LLC

© 2023 Hidenori Takahashi

Library of Congress Cataloging-in-Publication Data
Names: Takahashi, Hidenori, author.
Title: Ground improvement for coastal engineering / Hidenori Takahashi.
Description: First edition. | Boca Raton: CRC Press, [2023] |
Includes bibliographical references and index.
Identifiers: LCCN 2022058030 | ISBN 9781032211718 (hbk) |
ISBN 9781032211732 (pbk) | ISBN 9781003267119 (ebk)
Subjects: LCSH: Shore protection. | Soil stabilization. |
Soil compaction. | Grouting (Soil stabilization) | Soil amendments
Classification: LCC TC330 .T35 2023 | DDC 627/.58—dc23/eng/20230111
LC record available at https://lccn.loc.gov/2022058030

ISBN: 978-1-032-21171-8 (hbk)
ISBN: 978-1-032-21173-2 (pbk)
ISBN: 978-1-003-26711-9 (ebk)

DOI: 10.1201/9781003267119

Typeset in Sabon
by Newgen Publishing UK

Contents

3   Ground Improvement Methods   47

Preface

Since ancient times, coastal areas have been used as places for human life and habitation. They have been developed into hubs for the intersection of marine and land transport and mooring sites for fishing boats as well as cargo and passenger ships. Industries that transport raw materials and products by ship have also been developed along coastal areas. Airports, currently the centres of air transport, have been built in coastal areas because of the availability of wide flat lands. Furthermore, vast reclaimed land has been used for recreational areas, green spaces, housing, commercial premises, and offices not only as places for daily life but also as venues for important socio-economic activities. However, soft soils are typically deposited in coastal areas. Hence, measures to resolve problems associated with soft soils when constructing civil engineering structures, such as breakwaters, quays, revetments, reclaimed land, roads, bridges, and tunnels, must be implemented. In the medieval and early modern periods, when the economy began to develop, structures were built only in so-called 'good harbours', avoiding areas where soft ground was deposited. In modern and contemporary times, because structures are large and land is scarce, the use of improved soft ground has become a necessity.

Several books on ground improvement methods have been published [1, 2]. However, all of them are mainly concerned on ground improvement on land; none has focused on ground improvement at sea or in coastal areas. Densely populated coastal areas are often located near estuaries with soft ground. High groundwater levels also cause liquefaction problems in soft sandy soils. Hence, ground improvement in coastal areas is necessary. Accordingly, this book focuses on ground improvements in coastal areas. In the first half of this book, Chapter 1 provides an overview, and Chapter 2 discusses the general sequence of investigation, design, and execution. Books introducing the details of each ground improvement method already exist, and such details have been compiled in design and execution manuals. However, to the best knowledge of the author, no book explains how to proceed with investigation and design and how to select a ground improvement method when constructing a civil engineering structure. The second half of the book (Chapters 3–5) outlines the various ground improvement methods used in coastal areas. Chapter 3 describes the methods for improving the in situ ground, Chapter 4 describes the methods for utilising the improved soil, and Chapter 5 elaborates on the measures for the ground containing harmful substances. Note that this book is mainly intended for readers who require knowledge of the overall picture of ground improvement. Thus, the description of each technique is limited to a summary of the method and an outline of its investigation, design, and execution. Gaining a fundamental understanding of the ground improvement system is the first step. Then, an individual can deepen the understanding by acquiring further knowledge. After reading this book, if you find that knowing the details of each method is necessary, referring to books or manuals that centre on explaining the method is advantageous. In view of the large number of ground improvement methods and the complexity of situations in each country, a dictionary-like book covering the details of all ground improvement methods is difficult to produce; in fact, such a book may not be necessary. Accordingly, the details of ground improvement methods are left for other books to discuss.

I wish to briefly introduce myself and my motivation for writing this book. During my student days, I was affiliated with a laboratory in the field of coastal engineering where I processed data on observed waveforms in oceans. When I started working in my current place of employment (Port and Airport Research Institute (PARI)), I joined a laboratory in the field of geotechnical engineering specialising in ground improvement. Although this is a completely a different field of research compared with that during from my student days, it is interesting, considering the complexity of the topic. After completing my research on ground improvement and obtaining a PhD degree, I worked for two years at the Ministry of Land, Infrastructure, Transport, and Tourism (MLIT) in a department responsible for the design of port and airport facilities in the metropolitan area of Japan. Here, I obtained a professional engineering qualification in Japan (P.E. Jp) and gained considerable knowledge regarding site investigation, design, and execution. In performing design based on preliminary investigations, I learned the importance of considering whether ground improvement is necessary. If ground improvement is required, which method must be selected before deciding on it details? This was one of the motivations behind writing this book. Although books describing the details of each ground improvement method have been published, no book explains the procedures for designing ground improvement in the first place—a problem I encountered while working in the department where I was assigned. I have written this book not only as a memorandum for myself but also with the hope that it will benefit future generations.

After gaining experience in MLIT, I returned to PARI and worked on my research. This time, I was assigned to a laboratory working on the composite behaviours of waves and ground rather than on ground improvement. I started to study the destabilisation phenomena of coastal structures in response to tsunamis and high waves. This research project also provided me with an opportunity conduct international research in London. Eventually, I realised that ground improvement is typically necessary to increase the stability of structures subjected to waves. Since then, I have been the head of the Soil Stabilization Group, conducting research on ground improvement in coastal areas, advising on technical issues in the field, writing design criteria for port facilities, and disseminating these criteria abroad. Overseas dissemination is another motivation for writing this book. In Japan, cities are built in the estuaries of coastal areas, and large populations are concentrated in these cities where the ground is typically soft. In addition, earthquakes frequently occur in Japan, and problems related to seismic forces and liquefaction are encountered. Against this background, many ground improvement methods have been developed and implemented, setting Japan a leading country in the field of ground improvement. However, books introducing Japanese ground improvement methods are far fewer worldwide. The motivation for writing this book was to make people aware of the technology and encourage them to use it. Accordingly, this book introduces many of the methods used in Japan, including case studies with some regional bias. The design standards for port facilities [3, 4] are also useful for Japanese technology.

The author thanks those who assisted in writing this manuscript. Much of the information in this book has been obtained from previous books and standards. First, I wish to thank the technical developers and those who compiled the data. Experienced practitioners checked the content of each method. I wish to express my gratitude to Mr. Yuki Imai of Fudo Tetra Corporation, Mr. Sachihiko Tokunaga of the CDM Association, Dr. Ayato Tsutsumi of the Penta-Ocean Construction Company, and Mr. Hiroyuki Saegusa of Toa Corporation for their cooperation. I would also like to thank my former supervisor, Prof. Masaki Kitazume, who taught me the fundamentals of ground improvement and encouraged me to write this book. Finally, the author's laboratory secretary, Ms. Kanade Yagi, was instrumental in checking the manuscript and helping with the drawings for this book. The author expresses gratitude to all those who provided assistance.

Yokohama
December 2022
Hidenori Takahashi

REFERENCES

1. Kirsch, K. and Bell, A. 2013. *Ground Improvement*. CRC Press.
2. Han, J. 2015. *Principles and Practice of Ground Improvement*. Wiley.
3. Ports and Harbours Bureau, Ministry of Land, Infrastructure, Transport and Tourism (MLIT). 2018. *Technical Standards and Commentaries for Port and Harbour Facilities*. The Ports & Harbours Association of Japan (in Japanese).
4. Ports and Harbours Bureau, Ministry of Land, Infrastructure, Transport and Tourism (MLIT), National Institute for Land and Infrastructure Management (NILIM), and Port and Airport Research Institute (PARI). 2020. *Technical Standards and Commentaries for Port and Harbour Facilities in Japan*. Overseas Coastal Area Development Institute of Japan.

Author

Hidenori Takahashi graduated from Kyoto University in 2000, obtained his Master of Engineering in 2002. He then joined the Port and Airport Research Institute (PARI) as a researcher of geotechnical engineering. In 2008, he received a PhD from Kyoto University on the behaviour of ground improved by Sand Compaction Pile method. He is currently the head of Soil Stabilization Group of PARI, after working for the Ministry of Land, Infrastructure, Transport and Tourism in charge of designing port and airport facilities in the metropolitan area of Japan, and is a visiting researcher at Imperial College London.

He has been involved in the design of a number of sites and has contributed as a member of technical committees. He also has published many journal papers, mainly on the geotechnical aspects of ground improvement and composite behaviours of wave and ground. He was awarded three-time Best Paper Awards from Ports and Harbours Association of Japan in 2009, 2014, and 2018; four-time Best Paper Awards from Japanese Geotechnical Society in 2013, 2014, 2015, and 2021; Young Scientists' Prize from Ministry of Education, Culture, Sports, Science and Technology in 2017; Technical Development Award from Japanese Geotechnical Society in 2017; Best Paper of *Ground Improvement*, Telford Premium, ICE Publishing Award in 2019; and Best Paper of *International Journal of Physical Modelling in Geotechnics*, Telford Premium, ICE Publishing Award in 2020.

Chapter 1

Introduction

This chapter defines the soft ground in coastal areas and describes locations where soft ground can be found. The process of identifying soft ground is not merely based on specific soil parameters because it is influenced by the structure type and size. The chapter also presents the problems encountered when civil engineering structures are built on soft ground. Such problems involve stability and bearing capacity, settlement and deformations, liquefaction, and water-related problems. The chapter then describes how ground improvement has become necessary in coastal areas. Current ground improvement techniques are classified into five categories: replacement, consolidation and drainage, compaction, chemical treatment, and reinforcement; the principles involved in each type are explained. Finally, the ground improvement process is elaborated using a flowchart. The importance of construction and measurement as well as investigation and design is discussed.

1.1 SOFT SOIL IN COASTAL AREAS

When civil engineering structures, such as breakwaters and quay walls, are constructed in coastal areas, it is frequently necessary to strengthen the ground because the construction of structures on soft ground is difficult. In this chapter, soft ground in coastal areas is first defined; then, ground improvement as a countermeasure against the problems of soft ground is explained. Soft ground can be described as follows. When a coastal structure is to be constructed, the ground at the construction site may not be sufficiently stable to support the structure or resist external forces; consequently, the structure may fail. Even if it does not fail, the structure may not be able to achieve its intended function due to significant deformation during and after construction. This type of ground is defined as 'soft ground'. In general, clayey soils with high water content and highly organic soils are considered as soft soils; however, it is difficult to define soft ground in terms of specific fixed values. For example, a slight settlement in parks and green areas is not a severe problem, but an unacceptable inclination of the foundation of

DOI: 10.1201/9781003267119-1

a quay wall equipped with cranes is a dangerous problem. This means that ground softness is determined not only by ground conditions but also by the type and size of the structure to be constructed, the magnitude of external forces (e.g., seismic motion), the speed of construction, and the expected performance.

In coastal areas where soils are generally saturated with water, sandy soils may also be considered soft soils. In the absence of dynamic external forces, sandy soils are relatively stable. However, when shear forces due to vibrations, earthquakes, or waves act on the soil, liquefaction can occur, rendering the soil extremely soft. Liquefaction is a phenomenon that occurs when shear forces act on loosely deposited saturated sandy soil. These forces increase the pore water pressure and cause soil particles to lose their interlocking force; consequently, the soil behaves similar to liquid. In the past, earthquake-induced liquefaction has inflicted severe damage, such as the collapse of supporting structures, the displacement of structures under earth pressure (e.g., quays), and the flow of entire landfills. Hence, evaluating whether sandy soil is soft in terms of liquefaction is necessary. This includes assessing ground conditions, dynamic external forces (e.g., earthquakes), and whether the structure to be constructed is affected by liquefied ground.

Consider the soft ground found in coastal areas in detail (Figure 1.1). Many soft lands are found in these areas; they include deltas and spits near the mouths of rivers and alluvium formed by soft clay on the seabed. In addition, many tidal flats and wetlands are formed by tides and meandering rivers; soft sediments are deposited at the bottom of lagoons. The thickness

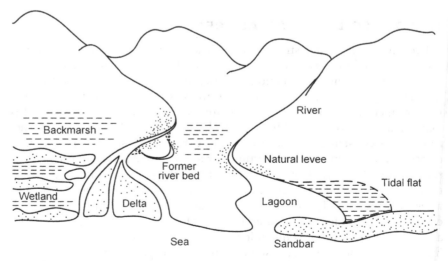

Figure 1.1 Soft soil in coastal areas.

and size of soft soil layer vary from place to place; however, only a limited number of sites are formed by solid bedrock. Moreover, most of coastal structures have to be constructed on the original soft ground.

In addition to the soft ground at the construction site, soft soils may also be used as landfill or backfill. In recent years, the necessity for large amounts of soil for large-scale reclamation projects has led to an increase in the use of soft soil considered as improved soil with satisfactory condition. A typical example of sediment material used in coastal structures is soil dredged to prevent navigation channels from being buried. In ports and harbours where breakwaters and quay walls are required, paths (i.e., shipping routes) are established for ships to pass through. To cope with the increasing size of ships, the water depth of the passage is deepened and often excavated below the surrounding seabed level. In this case, periodic dredging work is necessary because the sediments are deposited in the passage by water flow, and the water depth becomes shallower. The dredged soil is typically soft in composition because it is deposited on the surface layer of the seabed. Further, it has a high water content ratio because it is mixed with a considerable amount of water during dredging. Dredged materials are frequently used for coastal structures, such as ports and harbours, which are located in close proximity to the area where these materials are deposited. Hence, measures for dealing with soft sediments are necessary.

If no measures are implemented on the soft ground described thus far, the problems shown in Figure 1.2 can occur. If breakwaters and quay walls are built on soft ground, they may not be adequately supported and may collapse. Earthquakes may also cause the quay wall to lose its stability. Problems may not be observed during construction, but consolidation settlement may occur over a long period of time, inevitably causing deformation, which is dangerous to the structure. In addition, earthquakes can cause ground liquefaction, resulting in structure collapse or displacement. In coastal areas, the ground is constantly subjected to external hydraulic forces due to the inflow of water from land. This is in addition to storm surges as well as tidal waves, tsunamis, and ocean currents, which constantly change the pore water conditions. These cause seepage failure and scouring of soft ground. The foregoing problems associated with soft ground can be summarised as follows.

1. Stability and bearing capacity problems: ground failure, such as slope failure and foundation failure
2. Settlement and deformation problems: consolidation settlement and considerable ground deformation
3. Liquefaction problems: ground softening and sand boiling
4. Water-related problems: seepage failure and scouring caused by external hydraulic forces and pore water

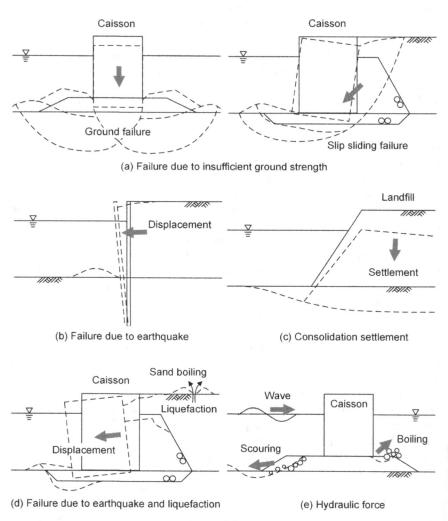

Figure 1.2 Problems associated with soft ground in coastal areas.

Because of the existence of these problems, it is necessary to determine whether the ground at the construction site of a coastal structure is soft or not. The next chapter discusses this in detail.

1.2 WHAT IS 'GROUND IMPROVEMENT FOR COASTAL ENGINEERING'?

In the previous section, soft ground in coastal areas was defined, and the problems associated with soft ground were summarised. Once the softness of the original ground has been determined, and the potential problems have been identified, the next step is to consider the countermeasures. One of these is to avoid soft ground in the first place and choosing a site with strong ground for the construction of a structure. Another alternative is to construct light and small structures, which do not experience stability and settlement problems even if the ground (e.g., clayey soil) is not sufficiently strong. The technique of driving piles into soft ground and securing the bearing capacity from them has also been widely used. The foregoing measures were common at the time when large-scale improvement of soft ground was not possible. For example, approximately 100 years ago in Yokohama, Japan, a large reclamation site was constructed to arrange quay walls, stockyards for imported and exported goods, and a terminal for passenger and transport railways. The reclamation was implemented according to the shape of the area where the hard clay layer was exposed to shallow depths (Figure 1.3). A few years after its construction, in 1923, due to the Great Kanto Earthquake, most of the quay walls collapsed. Nevertheless, the quay walls have been restored, and the artificial island remains in healthy use as a commercial and tourist area. As noted, another option is to construct light and small structures, as shown in Figure 1.4. Breakwaters made of stones piled up offshore, blocks of stones or bricks piled up to form quays or seawalls, or pile-type piers can be built to berth relatively large ships. Stone breakwaters are less likely to be subjected to large forces, and small block quay walls can be built on clayey soil if the load is distributed by stone mounds. These methods, however, are worth considering to avoid costly and time-consuming ground improvement.

With the modernisation of industry, structures have become increasingly larger, and solid ground on which structures can be built in coastal areas have become less, rendering the above methods infeasible. For this reason, large-scale ground improvement, which began a hundred years ago, has been implemented. In the field of civil engineering, ground improvement is fundamentally the improvement of soft ground to satisfy the function of the structure to be constructed. If a structure is built on soft ground, it may not be able to support the structure, or it may considerably deform after construction; evidently, ground improvement is necessary. As mentioned above, the use of soft soil and sand as improved soil and sand, respectively, with satisfactory condition has increased. Soil dredged from navigation channels, construction residue from land tunnelling, and soil from collapsed mountains and hills (for building artificial islands) are used in coastal area construction sites. The foregoing can be considered as a type of ground improvement. In particular, the use of dredged soil in its original state for the construction of structures is difficult because of its softness; hence, its improvement, either by dewatering or cementing, is necessary. Accordingly, in the scope of this book, the effective use of soft ground is considered as a type of ground improvement. Ground improvement has also attracted interest as a waste containment technology. Waste is used as a

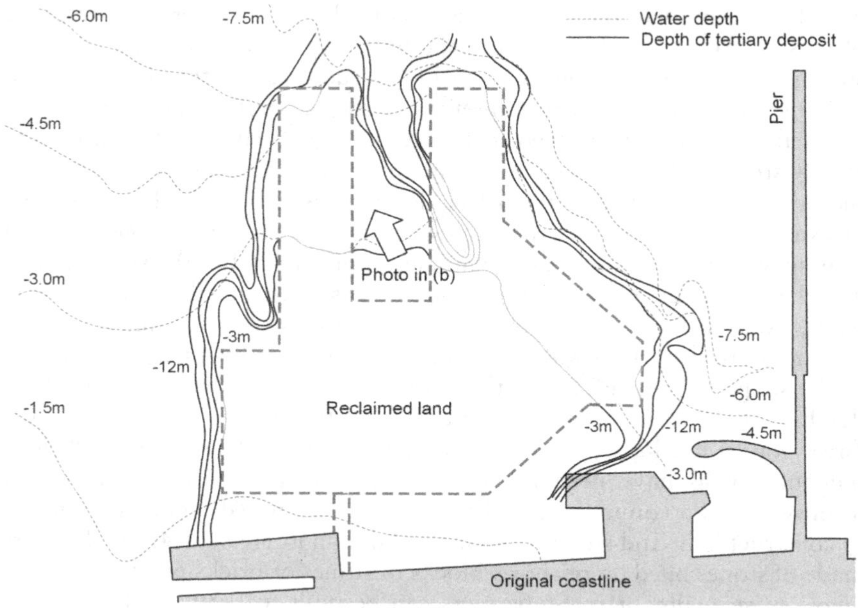

-6.0m   -7.5m   ······· Water depth
                                        —— Depth of tertiary deposit

-4.5m   Pier

                              Photo in (b)

-3.0m

                        -3m   -7.5m
              -12m
                                                                   -6.0m
-1.5m   Reclaimed land
                                        -3m   -12m   -4.5m

                                                      -3.0m

                              Original coastline

(a) Reclamation on hard ground

(b) Quays still in use

Figure 1.3 Artificial island placed on hard clay layer. (Based on Ref. [1].)

landfill material in coastal areas, and the permeability of the ground around wastes must be typically reduced to prevent contaminants from leaking into the surrounding ground and sea. This differs from soft ground improvement described thus far. However, the common point is that the function of the

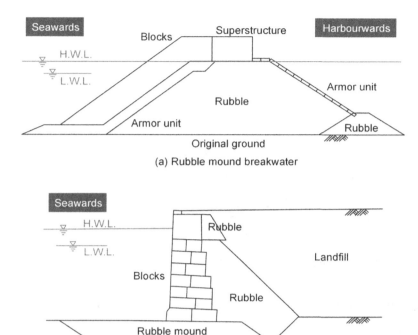

Figure 1.4 Light and small structures on unimproved ground.

structure is satisfied by modifying the ground, and this is treated as a type of ground improvement in this book.

In the last 50 years, ground improvement technology has considerably advanced, and various improvement principles and techniques have been developed and applied. This has resulted from the development of geotechnical engineering as a discipline, the improvement of construction machinery, and the formulation of new materials. As a background, the number of industries requiring ground improvement is increasing worldwide. Moreover, new and diverse requisites are arising. These include ground improvement around existing structures and the necessity for achieving the early effectiveness of ground improvement. Figure 1.5 shows an example of ground improvement in a coastal area; currently, the improvement of seabed and landfill sites for ports and airports has become a common practice. Ground improvements are implemented in many locations, such as the original ground of breakwaters, quay walls, and seawalls; the original and reclaimed ground of landfill sites; and roads and bridges in port areas.

Figure 1.5 Ground improvements in coastal engineering.

1.3 CLASSIFICATION OF GROUND IMPROVEMENTS

The principles of in situ ground improvement can be classified into four categories: (i) replacement, (ii) consolidation and drainage, (iii) compaction, and (iv) chemical treatment. The ground improvement principles in terms of the effective use of soft ground can be classified into three categories: (i) consolidation and drainage, (ii) chemical treatment, and (iii) reinforcement. Table 1.1 summarises the classification of various ground improvement methods based on each principle. Each ground improvement method is not necessarily based on one principle but frequently on several fundamentals. In addition, the manner of comprehending the principle depends on the ground and method of application. For this reason, some methods are described in more than one place in the table. Furthermore, in an actual site, several methods are often applied. The problem of soft ground at the site of a structure must be assessed, and an appropriate ground improvement method based on relevant principles must be selected. In some cases, a combination of principles and methods must be applied to deal with soft ground. The details of each method are explained in Chapter 3 onwards; in this chapter, each ground improvement principle is explained.

'Replacement' is one of the ground improvement methods that has been used since the early 1900s. In this technique, problematic soils are replaced by non-problematic soils. Initially, the method was applied by heaping excellent quality soil on top of soft clayey soil; the soil is replaced according to the weight of the fill. Subsequently, as construction machines became larger and more sophisticated, the excavation of soft ground (bed excavation) and

Table 1.1 Classification of ground improvement methods

(a) In situ ground improvement;

Principle	Name	Method
Replacement	Replacement	Replace with good quality soil
Consolidation and drainage	Vertical drainage (SD, PVD)	Place vertical drains to promote consolidation
	Vibro-replacement	Place stone piles to promote consolidation
	Vacuum consolidation	Reduce water pressure in drains to promote consolidation
Compaction	Sand compaction pile	Press-in sand piles to compact ground
	Vibro-compaction	Form gravel piles with vibration
	Heavy tamping	Drop weight to compact ground surface
	Compaction grouting	Inject grout and compact ground
Chemical treatment	Shallow mixing	Solidify shallow layer with cement
	Deep mixing	Solidify deep layer with cement
	Jet grouting	Mix soil with cement using high-pressure jet
	Chemical grouting	Infiltrate special chemical solution

(b) Use of improved ground;

Principle	Name	Method
Consolidation and drainage	Filter press	Dehydrate under high pressure
Chemical treatment	Pre-mixing	Pre-mix and solidify with cement
	Pre-mixing with secondary material	Pre-mix and solidify with secondary material such as slag
	Pneumatic flow mixing	Mix with cement in pipe
	Lightweight mixing	Pre-mix and solidify with cement and air foam
	Deep mixing	Solidify with cement using deep mixing
Reinforcement	Reinforcement with geosynthetics	Compound soil and geosynthetic

the placement of soil on the excavated area became the mainstream method. This is because excavation offers the advantage that the cross section can be built according to the design. However, the use of replacement is decreasing due to the problems of disposal of excavated soil and the liquefaction of replaced soil during earthquakes.

'Consolidation and drainage' method is the preferred approach for con-solidating cohesive soils. The consolidation of cohesive soil with low per-meability requires considerable time; moreover, the soil strength is low. To promote consolidation, drains are usually installed in the ground to quickly drain pore water from the clay. Because consolidation is not only achieved by embedding the drainage material into the ground, the drainage material must be used in combination with other methods, such as constructing an embankment on the ground surface where the top load can be applied. Negative water pressure can also be applied to the drainage material to absorb pore water. A number of drainage materials have been proposed and applied in practice, such as sand, crushed stone, artificial fibres, and plastics. Although some limitations exist (e.g., the necessity for controlling consoli-dation and the fact that the strength of the ground does not immediately increase after ground improvement), the method is relatively economical and has been actively used since 1950.

'Compaction' is a method for promoting the consolidation of cohesive and loose sandy soils. As mentioned, unconsolidated clay soils are soft and must be drained when loaded. Sand piles are formed on the ground, and vibrations are applied to the piles to increase their diameter as well as load the clay horizontally. The sand piles act as drains while applying load to the clay soil. Loose sandy soils are densified by dropping weights on the ground surface, inserting and vibrating a steel pile, or driving piles of crushed stone and vibrating them. The sand piles and grout masses are spread on the ground to densify the sandy soil and increase the horizontal confining pressure. This is because the stiffness and strength of sandy soils depend not only on their density but also on the confining pressure. These methods have been in use since 1960.

'Chemical treatment' is a technique that involves the addition of soil stabilisers, such as lime, cement, or chemicals, thereby modifying the soil through chemical reactions. Special chemical solutions that can penetrate the ground and then solidify or gel have been used since the 1960s. Another method is to embed a rod with an agitator into the soft ground and then mix and agitate cement into the ground to form a long cement-treated soil pile for ground improvement. The piles can be overlapped in succession to form a solidified wall or block in the ground. This method has rapidly become widely used since the 1970s. In addition to physical agitation using a blade, soil and cement agitation is also accomplished by injecting a jet of high-pressure cement slurry into the ground and moving it. The ground strength is immediately increased after construction. Moreover, the size of the con-struction machine can be reduced; the versatility of this method is high.

'Reinforcement' is not a method that modifies soil properties; it is a tech-nique that involves embedding a material that differs from the soil and enab-ling the material to share in resisting large forces (compression, tension, shear, etc.). The inadequate or missing soil properties can be complemented by the material to improve the overall performance. This method had been in long use; the original reinforcements were natural materials, such as wood. In recent years, artificial materials, collectively known as geosynthetics, have been generally employed as reinforcing materials. Geosynthetics, which include geotextiles (e.g., woven and non-woven fabrics, geogrids, and geonets) are lattices made of plastic; those manufactured in sheet form are called geomembranes. Although these methods and materials are widely used in terrestrial areas, their application to coastal areas remains limited.

1.4 PROCEDURE OF GROUND IMPROVEMENT

This section briefly explains the procedure in applying ground improvement to the construction of coastal structures. In the construction of a structure, the project undergoes various investigations, planning, design, execution, and maintenance. The following is a step-by-step explanation of how ground improvement is related to this process. Figure 1.6 shows the flow of processes in the entire project, including the consideration of ground improvement. The figure describes an exemplary flow of construction; in practice, parts of the flow may be omitted, the flow order may move back and forth, or supplementary investigations may be implemented. The general information applicable to all ground improvement methods is presented in Chapter 2, and the design and construction aspects specific to each method are presented in Chapter 3 and subsequent chapters.

Initially, information on the original ground of the construction site and the ground to be used for reclamation must be collected. In urban areas, the results of previous soil investigations may be available in a database and are useful to study the topography and origin of the land. An approximate geotechnical investigation must be performed although the number of borings may be small. Based on these investigations, the function, characteristics, and importance of the structure are determined; then, a schematic design is implemented on the assumption that no ground improvement is necessary. This design must comprehensively consider safety during construction, construction period, stability and deformation after construction, and stability under accidental external forces (e.g., earthquakes). If the assessment indicates that the function of the structure cannot be maintained, such function must be re-evaluated, or ground improvement must be implemented. Because ground improvement works generally require considerable budget, considering the option of avoiding ground improvement by reviewing some of the functions of the structure is worthwhile. In contrast, there are cases where the necessity for ground improvement is evident despite the absence of a schematic design. For example, construction of a gravity-type structure on extremely soft clay soil is impossible. In such a case, the approximate design on the original ground may be omitted, and the study may proceed on the assumption that the ground is to be improved.

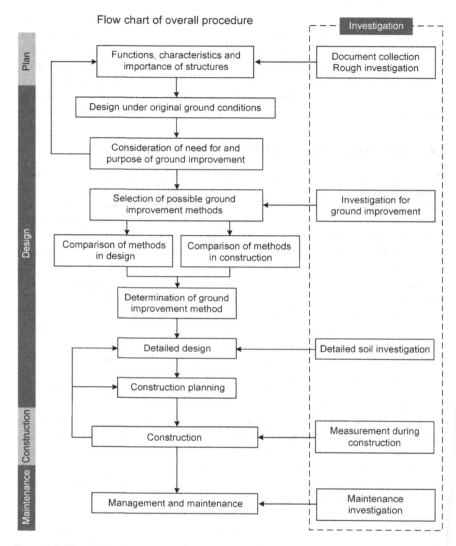

Figure 1.6 Flowchart of overall procedure.

When the decision is to implement ground improvement, several methods are selected considering the purpose of the improvement, the characteristics of the target soil, the construction period, and the influence on the surrounding area. Depending on the method chosen, additional investigations may be required. For example, if vibration and noise are concerns, then it is necessary to understand the facilities and constraints in the vicinity of the site. If the technique involves cement deep mixing, it is necessary to determine whether soil in which the application of cement is difficult to implement exists in the site. Based on the selected methods, design proceeds by considering the feasibility of execution; the perspectives of design and execution are compared. Finally, the most suitable method is selected based on its overall economic efficiency. Once the ground improvement method has been selected, full-scale ground investigation and detailed structure design are implemented. In design, not only the design of the structure but also the design of ground improvement must be implemented; the design of the whole structure must integrate both methods. After the design, a construction plan is devised. In this plan, the ground improvement procedure and the order of constructing the structure must be considered. For example, if several ground improvements are to be applied, an incorrect procedure can lead to significant displacements in existing structures. Hence, a construction plan is critical.

In the actual construction phase, monitoring ground and structural movements during construction is extremely important. The design is only a desk study and typically differs from the actual behaviour. In some cases, modifying the design or reviewing the construction plan may be required. Upon the successful construction of the structure, maintenance is the subsequent step. For example, if the structure was designed and constructed to allow consolidation settlement, the amount of settlement must be measured over time. Further, if the amount of settlement differs from the prediction, countermeasures must be formulated and implemented. Ideally, ground improvement must be provided to ensure that the structure performs satisfactorily throughout its service period. This is necessary because once the superstructure has been constructed, improving the ground becomes costly and labour-intensive. Moreover, proper maintenance and management in the years to come are not ensured and usually have no allotted budget. However, if achieving a long-term and maintenance-free ground improvement requires considerable labour and budget, then it must be assumed that maintenance and repairs are to be implemented throughout the structure's service life. The succeeding chapters provide further details of the procedure.

Reference

1. Temporary Customs Department. 1906. *Report on the burial of works on the sea at Yokohama Customs*. Temporary Customs Department.

Chapter 2

Investigation, Design, and Execution

This chapter describes the investigation, design, and execution of structures including ground improvement in coastal areas. Investigations are performed prior to conducting schematic and detailed design and during execution. In particular, the methods and frequency of geotechnical investigations are described. Further, this section outlines the items to be considered for the design of coastal structures and elucidates the solutions to problems caused by soft ground (stability and bearing capacity, settlement and deformation, liquefaction, and water-related problems) through ground improvement. It also introduces the actual design procedure for quay walls and demonstrates the incorporation of model tests and numerical analyses into the design. Finally, the execution, including the use of information technology and execution management methods, is presented.

2.1 INVESTIGATION

Chapter 1 explained that the conditions of the ground and surrounding facilities must be investigated at each project stage—from planning to maintenance. In addition, the investigation of other factors, such as construction constraints (e.g., vibration and noise) and the presence of underground structures, is also important for the construction of coastal structures. The purpose of geotechnical investigation is to understand the stratigraphic composition and the engineering properties of each layer of the ground. This chapter describes the general investigations applicable to all ground improvement methods, and Chapter 3 onwards elaborates on the specific investigations required for each ground improvement method.

The investigations have been implemented according to the order of overall project flow, as shown in Figure 1.6 (Chapter 1). In the planning stage, information from literature, existing documents, and soil databases is collected. A preliminary investigation is implemented to assess the situation at the construction site. At this stage, comprehending the problems at the site based on collected data and determining the items and quantities to

DOI: 10.1201/9781003267119-2

be investigated in the field are preferred. For example, if the stratum composition is complex, then the number of investigations must be increased; if the soil is viscous, then sampling must be conducted to determine the mechanical properties; and if a history of previous studies exists, then the number of investigations and items to be examined may be reduced. In the case of coastal investigation, the number of items to be scrutinised may be decreased. Coastal investigations differ from terrestrial investigations in many respects. First, coastal structures, such as breakwaters and landfill sites, are usually long, and the scope of investigation required for design is broad. Furthermore, ground investigations conducted at construction sites require considerable time and effort. They require tasks, such as the preparation of offshore oars, as shown in Figure 2.1, because the investigations are performed at sea. In contrast, understanding the stratigraphic composition is generally easier because the site consists of sediment deposits that are relatively new and undisturbed by human activities. For these reasons, in addition to boreholes and soundings, increasing the measurement intervals and implementing geophysical exploration to obtain a three-dimensional (3D) aerial view of the ground are necessary in some cases especially in the early stages of the outlined investigations.

If the implementation of ground improvement is decided based on preliminary design calculations, a number of candidate ground improvement methods must be identified, and further studies must be implemented to determine the applicability of each method. The special investigations required for each method are described in Chapter 3 onwards. As mentioned above, ground investigations in coastal areas are time-consuming and costly. Therefore, investigations for ground improvement are typically included in the outline of investigations or in the main investigation described below. Once a ground improvement method is selected, a full-scale investigation

Figure 2.1 Soil investigation at sea. (Courtesy of Kiso-Jiban Consultants Co.)

is conducted for a detailed design. In the main investigation, the number of investigation methods is determined based on understanding the design and construction problems of the structure. If bearing capacity and consolidation are problems, then the soil must be sampled to determine its shear properties, and consolidation tests must be conducted, respectively. If liquefaction is a concern, then grain size and in situ soil strength must be determined. In addition to ground investigation, scrutinising existing conditions is necessary for ground improvement. Conditions, such as the area of the site, the terrain, the delivery route of construction equipment and materials, the source and route of materials for the main and temporary works, existing structures, facilities around the site, permissible vibration and noise, fluctuation and flow of groundwater, water quality, and status of rare species, must be examined. These investigations are useful for the design and implementation of ground improvement methods.

The number of ground investigations for each construction site must be appropriately evaluated. However, less experienced engineers may require an approximate guide. As a reference, Table 2.1 lists the general spacing of investigation points. In the table, the intervals are classified into normal and perpendicular directions, that is, the longitudinal and perpendicular directions of breakwater or quay, respectively. The investigation must cover an area that is 50–100 m from the breakwater or quay. For coastal structures that are not linear (such as breakwaters and quays) but are planar (such as landfill sites), the intervals in the column marked 'normal direction' must be used as reference. In addition, the term 'sounding' refers to

Table 2.1 Approximate spacing of investigation points

(a) When the ground is horizontally and vertically homogeneous;

		Normal direction		Perpendicular to normal	
		Boring (m)	Sounding (m)	Boring (m)	Sounding (m)
Preliminary investigation	Wide area	300–500	100–300	50	25
	Small area	50–100	20–50		
Detailed investigation		50–100	20–50	20–30	10–15

(b) When the ground is not homogeneous

	Normal direction		Perpendicular to normal	
	Boring (m)	Sounding (m)	Boring (m)	Sounding (m)
Preliminary investigation	less than 50	15–20	20–30	10–15
Detailed investigation	10–30	5–10	10–20	5–10

Source: Based on Ref. [1].

an investigation that does not require boreholes (e.g., cone penetration test (CPT)). If a standard penetration test (SPT) is to be conducted, then the reference interval is marked 'boring'. The intervals are provided only as guides, and the number of investigations must be appropriately modified. Moreover, the size, type, weight, foundation type, construction cost, importance of the coastal structure to be built, and the stratification and homogeneity of the ground must be considered. The depth of investigation is usually limited to the strata with sufficient bearing capacity. For relatively small and large structures, the N values must at least be 30 and 50, respectively. For earthquake-resistant facilities, the depth of investigation must consider a shear wave velocity of 300 m/s (a base value used in engineering).

Typical ground investigation items are selected, as listed in Table 2.2. The role of each investigation varies because the preliminary investigation provides an approximation of the local conditions. In contrast, the main investigation is necessary to obtain information for detailed design and construction studies. The main purpose of the preliminary investigation is to derive information on the composition of soil layers as well as the types and properties of the soil, whereas the objective of the main investigation is to obtain information on mechanical properties. The ground investigation items must be implemented according to the foregoing objectives. In Table 2.2, the items to be considered from the preliminary investigation to the main investigation are classified into four categories: (1) effective to implement; (2) implemented after comparison with other methods; (3) implemented when necessary; and (4) not usually implemented. However, these categories must be considered simply as guides because the perspectives of different countries and standards on investigation methods vary. For example, in Japan, the SPT is usually employed for in situ investigation and testing; when ground conditions are predictable, the CPT is more common. The standards of each country and previous cases must be examined to determine the ground investigation items to be included. In recent years, the limit state design method based on reliability theory, which considers the variation of various parameters in the design, has been introduced worldwide. This method (described in the next section) is changing the choice of the items to be included in ground investigations. The more the investigation items, the more expensive the investigation. However, the investigation becomes more accurate, and the ground strength used for the design can be better estimated. Cost savings are usually made on investigations, which are frequently regarded as wasteful; however, cutting on their costs can result in over-built structures and higher overall construction costs. Hence, developing a strategy to decide what to investigate is important.

In situ investigation and testing can be divided into two main categories: boring with associated tests and sounding (a method not requiring a borehole). Ground hole boring provides information on the composition of soil layers and groundwater level at that location. The holes can also be

Table 2.2 General items for ground investigation

Classification	Investigation items	Stability Clay	Stability Sand	Settlement/ deformation	Permeability	Dynamic and liquefaction properties
In situ investigations and tests	Boring	1	1	1	1	1
	Sampling	–	3	–	3	3
	Standard penetration test (SPT)	2	1	2	1	1
	Permeability test	4	4	3	1	4
	Seismic velocity logging (PS)	4	4	4	4	1
	In-hole load test	3	3	3	4	4
	Sounding	3	3	3	4	3
Laboratory tests — Physical properties	Water content	1	3	1	4	4
	Liquid and plastic limits	1	3	–	4	1
	Soil particle density	–	3	–	3	3
	Soil particle size	–	3	–	3	–
	Wet density	–	3	–	4	3
Mechanical properties	Unconfined compression test	1	4	2	4	4
	Triaxial compression test (UU/CU/CD)	2	2	2	4	4
	Consolidation test	4	4	1	4	4
	Permeability test	4	4	4	3	4
	Cyclic undrained triaxial test	4	4	4	4	3
	Cyclic triaxial test	4	4	4	4	3

Note: 1 = Effective to implement; 2 = implemented after comparison with other methods; 3 = implemented when necessary; 4 = not usually implemented.

used to obtain soil samples for various tests (e.g., SPTs, in situ permeability tests, and velocity logging). Soundings that do not require boreholes include the cone penetration, Swedish sounding, and vane shear tests. In these tests, a rod with a cone or blade attached is driven into the ground; the penetration resistance and rotational torque are measured to determine the ground strength. Although each method has its own advantages and disadvantages, the use of the CPT is increasing worldwide because it is more economical than the SPT. However, the latter enables the collection of samples. Hence, the implementation of in situ investigation and testing method based on understanding the characteristics of both methods (such as using the SPT with the CPT or when ground conditions are not well known) is advisable. For the definitions and methods of in situ investigation and testing, the reader is referred to textbooks on soil mechanics and ground investigation standards of various countries. For example, British Standards (BS), American Society for Testing and Materials (ASTM), and Japanese Geotechnical Society (JGS) standards are useful as they have been in use for a long time and carefully revised.

In situ investigations cannot provide detailed information on soil properties; hence, soil samples are brought to the laboratory for testing. There are two main soil sampling methods: soil sampling without preserving the sedimentary structure (disturbed sampling) and soil sampling without disturbing the sedimentary structure (undisturbed sampling). Because the collection of soil samples with the same sedimentary structure and condition is virtually impossible, the use of the expression 'less disturbed' instead of 'undisturbed' seems more appropriate. A disturbed sample is essentially a specimen extracted by an SPT sampler from which physical properties (e.g., water content, density, and grain size) and chemical properties (e.g., pH and organic matter content) can be determined. Samplers suitable for the type of soil are employed to obtain undisturbed samples. Thin-walled samplers are used for soft clayey soils, whereas double-tube or triple-tube samplers are used for collapsible and fragile clayey or silty soils. The sampling of gravelly soil requires more sophisticated techniques, such as freezing or gel sampling. Note that many sampling methods have been proposed worldwide. Consequently, the extraction of soil samples with the same accuracy is not always possible even if the sampling method is claimed to be equally effective. The characteristics of sampling methods can be determined by examining literature and results that have been published worldwide. The use of inexpensive sampling methods that are claimed to be of high quality may result in uneconomical designs because the soil sample strength may be underestimated due to high variability. Various mechanical properties (e.g., shear, consolidation, permeability, and dynamic properties) can be determined by performing mechanical tests on the collected samples. For the definitions and methods of soil sampling and laboratory tests, textbooks on soil mechanics and national standards for soil testing (e.g., BS, ASTM, and JGS standards) must be consulted. In recent years, the construction of coastal structures has increasingly become globalised with companies from other countries often collaborating with design and construction teams. Even if the objectives of determining the physical and mechanical properties of soils are the same, the test methods may differ; hence, checking the content of the standards specified for the project under consideration is required. Further, examining methods for evaluating the results of tests implemented according to these standards is necessary.

2.2 DESIGN CONCEPT

Once the overall project plan (e.g., structure function, characteristics, and importance), information from sources (e.g., literature, existing documents, and soil databases), and preliminary investigation have been completed, an approximate design is implemented with the assumption that ground improvement is not implemented. Even if ground improvement is required, understanding the problems of unimproved ground and clarifying the expected effects of ground improvement to select and design an appropriate ground improvement method are required. If ground improvement is clearly required, the approximate design of the unimproved ground may be omitted. Then, a design based on the assumption that ground improvement is to be implemented is immediately conducted. However, identifying the problems of unimproved ground first (even without conducting design calculations) is preferable. If it is deemed that the function of the structure cannot be maintained in the unimproved ground, then the function is reviewed, or the implementation of ground improvement is considered. If the decision is to implement ground improvement, several methods are selected. Specific design calculations are also conducted considering the characteristics of the structure and target soil, purpose and characteristics of ground improvement, construction period, and impact on the surrounding area; further details on the foregoing are described below. The design and construction aspects are compared, and the most suitable method is selected based on the overall economic efficiency. Once the method is determined, the main investigation is implemented to develop a detailed design and construction plan for the structure. The design involves confirming the design methods of the structure and ground improvement; hence, both methods must be integrated.

The characteristics of the structure and target soil, purpose and characteristics of ground improvement, construction period, and impact on the surrounding area are the items to be considered. The construction method must be selected based on a comprehensive evaluation. However, this is easier said than done; a high level of skill and experience for each item is necessary. Hence, referring to this book and similar design documents in the past is advisable. The comments on each item are as follows.

1. Structural characteristics

   No structure is insusceptible to failure; however, the impact of the failure of certain structures may either be immediate and severe or less extreme. The tolerance for the amount of deformation also varies. For example, a seawall in a landfill does not lose its function even if it slightly deforms, whereas a crane foundation (for unloading cargo) built on the same landfill is sensitive to deformation. The characteristics of such structures are a major controlling factor when determining the extent and scale of ground improvement. The conditions of the structure are frequently specified in the form of safety factors against failure, allowable settlement, and allowable slope.

2. Properties of target soil

   The properties of the target soil are evidently of paramount importance in the design of ground improvement. The improvement principle of each method is typically limited by soil characteristics. For example, consolidation methods are ineffective in sandy soils. The stratification conditions of the soil are also factors that cannot be ignored in the method selection. For instance, when several soft clay layers with different consolidation coefficients are interlayered, the rate of consolidation greatly varies depending on the stratigraphic order of the layers; this is even more evident in the case of permeable layers.

3. Purpose of ground improvement

   The problem with the original ground, the purpose of ground improvement, and the degree of ground improvement must be clarified. For example, even if settlement is a problem, the preload size and loading period may vary depending on the allowable residual settlement. In extreme cases, the use of foundation structures may be necessary to sustain settlement beyond the limit that ground improvement can provide.

4. Characteristics of ground improvement methods

   To select a suitable method from possible ground improvement techniques, understanding the characteristics of each method is necessary. The following items must be considered in the selection of ground improvement method: (a) design accuracy; (b) design cost (accuracy and quantity of required preliminary investigation and analysis method); (c) construction capacity (possible depth and loss due to construction scale); (d) construction limit in the field; (e) quality and quantity of materials, machines, power, and manpower; (f) construction certainty (difference among accumulated construction results); and (g) degree of difficulty in modifying design (change in construction plan) in case of construction problems.

5. Construction period and impact on surroundings

   The period allowed for construction is one of the major factors limiting the choice of ground improvement methods. Regardless of how small the cost, a method that requires an extremely long time before targets can be achieved is impossible to implement. Thus, if the construction period is short, then the vertical drainage method may be selected instead of the pre-loading method, or if the work completion in the site is more urgent, then the cement improvement method may be opted. Environmental considerations must also be taken into account. It must be ensured that existing structures adjacent to the site are not adversely affected and that noise and vibration are not excessive. As the conditions require, measures may have to be implemented to reduce the impact on the surrounding area; in certain cases, the choice of construction method may have to be reconsidered.

Specific design calculations are conducted to determine whether problems associated with soft ground (described in Chapter 1) may be encountered. If such is the case, then a course of action, such as the implementation of ground improvement, must be implemented to prevent the problem from occurring and to achieve the function of the structure to be constructed. The problems associated with soft ground can be restated in terms of the following four categories.

1. Stability and bearing capacity problems: ground failure, such as slope failure and foundation failure
2. Settlement and deformation problems: consolidation settlement and considerable ground deformation
3. Liquefaction problems: ground softening and sand boiling
4. Water-related problems: seepage failure and scouring caused by external hydraulic forces and pore water

The fundamental concept of design calculations for unimproved ground and ground improvement is explained in the succeeding section. In this section, the current mainstream design methods, that is, the performance design method and the limit state design method, are discussed.

In the past, the concept of specification design (i.e., design was implemented according to specifications in design standards) was frequently applied. The safety and performance of the structure were determined by the author of the standard, and the stability of the structure was ensured, provided that the design was conducted in accordance with the standard. Specifically, the allowable stress design method was used. The safety factor was specified in the design standard, and the design was implemented by ensuring that this factor was satisfied. However, currently, the concept of performance design in lieu of specification design is used in many design standards. Performance design is defined in various ways; however, in general, it is a method of specifying the required performance (e.g., repairability and serviceability) of soil or a structure and then designing it to satisfy performance. Any design method can be used if the required performance is satisfied. The advantages of performance design include the high degree of design freedom and the clarity of objectives to be achieved. This freedom means that keeping abreast with technological progress and complying with international design standards are facilitated. Performance design also has its disadvantages. The high degree of freedom in design requires a high level of skill to examine and determine conformity; similarly, design requires considerable technical ability. In recent years, research on the non-linearity of materials and soil and studies on numerical analysis techniques, such as finite element analysis, have led to rapid developments. Consequently, design and examination using advanced techniques have become possible. In performance design, any specific design method may be employed; however, in practice, a method called limit state design is widely used. In the limit state design method, the failure mode (e.g., circular slip or foundation failure modes) is assumed when the ground or structure reaches the limit state (e.g., failure). The probability of failure is designed to be less than a certain level based on the reliability theory, considering the variability of loads and soil strength. This design method is preferred because it enables the use of the same failure and deformation modes as well as formulas used in the conventional allowable stress design method; consequently, the sudden variation in design methods is reduced.

The limit state design method probabilistically evaluates performance based on the reliability theory; however, in practice, a simple method is frequently applied. To verify the performance, the design values are calculated by multiplying the characteristic values and partial factors and confirming that the design value of the resistance force, R_d, exceeds the design value of the acting force, S_d. To determine the margin of safety, S_d/R_d (design value of the acting force divided by the design value of the resistance force) may be used as reference. It may also be multiplied by the structure factor, γ_i, given below to account for the social and economic effects of the structure, and designed to be less than 1.0 (Eq. (2.1)):

$$\gamma_i \frac{S_d}{R_d} = \gamma_i \frac{\gamma_a \gamma_s S_k}{\gamma_R R_k} \leq 1.0, \tag{2.1}$$

where γ_a is the structural analysis factor; γ_R and γ_S are the partial factors (i.e., resistance and load factors, respectively) of the resistance and acting forces, respectively; and R_k and S_k are the characteristic values of the resistance and acting forces, respectively. This technique is called Level 1 reliable design method or partial factor method. The characteristic values are representative of the statistically distributed values. The characteristic values of soil may be set using various means. For example, values may be derived by the first-order processing of measured values or by transforming them into different engineering quantities from empirical values; the average of the derived values is then used as the characteristic value. Alternatively, the derived value may be multiplied by a correction factor to consider variations depending on the reliability of data or ground investigation method or soil testing method. The resultant may be used as the characteristic value. In this case, the effect of these factors can be eliminated from the partial factors for each failure and deformation mode, thus simplifying the performance verification. The partial factors are intended to secure the margin of safety to account for individual uncertainties in the performance verification. The various types of partial factors to account for soil strength variations, load variations, and uncertainties in structural analysis include resistance, load, and structural analysis factors, respectively. In the field of geotechnical engineering, resistance and load factors are often specified, or the safety factor used in the allowable stress design method is replaced (such as by the structural analysis factor), assuming that the resistance and load factors are both 1.0. Note that the methods of calculating the characteristic values and setting the partial factors differ according to the country, standard, and time of publication of the standard. Thus, confirming the contents of the standard specified in the project under consideration is necessary.

2.3 DESIGN CALCULATION APPROACH

2.3.1 Stability and Bearing Capacity

Actual design calculations identify the problems of soft ground and provide solutions to these problems. Some examples of problems and solutions are discussed in the following sections. First, the most important concern associated with soft ground is the stability problem caused by inadequate soil strength. The slip failure of entire soils in seawalls and quays, earth pressure problems, and foundation failure due to piles and heavy loads are among the stability problems in the design of coastal structures.

Figure 2.2 shows an example of the stability problem of a seawall under a circular slip condition. If the resistance–load ratio (i.e., the ratio of the moment due to the force causing the soil mass to slide along the circular slip surface to the moment due to the force that resists it) is less than 1.0, then the seawall loses stability, and slide failure occurs. In most cases, the partial factor multiplied to each parameter is 1.0, and the overall structural safety analysis factor is 1.2–1.3; the partial factors may be also adjusted to achieve the same level of stability. When the resistance–load ratio is less than 1.0, the deformation close to failure (i.e., shear deformation, such as lateral flow and immediate settlement) becomes extremely large such that the function and performance of the structure are frequently not satisfied even if slip failure does not occur. In the past, numerous cases where the ground lost its stability and failed by sliding have occurred. Similar cases are observed today because of inadequate investigation, design, and construction. If slip failure is anticipated to occur in unimproved ground, the application of ground improvement is considered as one of the countermeasures. Sliding failure is caused by the lack of resistance to the sliding forces of soil mass. For example, as shown in Figure 2.2, the deep mixing method can be applied to increase the soil strength, thereby increasing the resistance–load

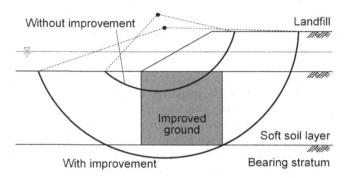

Figure 2.2 Stability study using circular slip analysis.

ratio. Note that designing a method to merely deal with a localised part of the slip surface by solidification is dangerous; solidification must be implemented throughout the area involved. This is because if the strength of only one part of the slip surface is increased, the slip surface may avoid it, or another type of slip surface may occur. A possible solution to this problem is the use of lightweight materials behind the seawall to reduce the sliding force of the soil mass.

In addition to the overall ground stability, inadequate resistance to earth pressure is another critical stability problem. In the case of sheet pile seawalls and quay walls, the high active earth pressure and low passive earth pressure acting on the sheet piles may result in the loss of structural stability. The active earth pressure, P_a, and the passive earth pressure, P_p, acting on a sheet pile in a clay layer can be expressed by the following equations according to a simplified model:

$$P_a = \frac{1}{2}\gamma H^2 K_a - 2cH\sqrt{K_a},\tag{2.2}$$

$$P_p = \frac{1}{2}\gamma H^2 K_p + 2cH\sqrt{K_p},\tag{2.3}$$

where γ is the weight per unit volume of soil; H is the height of the soil layer; K_a is the coefficient of active earth pressure; K_p is the coefficient of passive earth pressure; and c denotes the soil cohesion. In these equations, if the cohesion, c, is low owing to the ground softness, the magnitude of active earth pressure that can move the sheet pile seaward is high, whereas that of the passive earth pressure to support it is low. This indicates that the stability of the sheet pile is also low. If the sheet pile does not satisfy the specified stability, a countermeasure is considered, such as increasing its

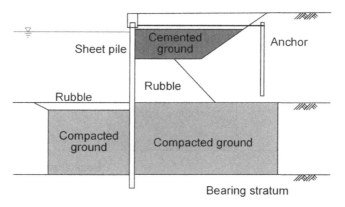

Figure 2.3 Sheet pile quay walls reinforced by ground improvement.

stiffness and strength, deepening the penetration into the supporting layer, or pulling the upper part of the sheet pile from the anchorages using tie rods. However, these methods have limitations; further, the sole use of sheet piles may be uneconomical. In this case, the application of ground improvement may be considered (Figure 2.3). Eq. (2.2) suggests that reducing the unit weight or increasing the cohesion reduces the active earth pressure. In addition, if the angle of internal friction of the soil can be increased, the coefficient of active earth pressure decreases, thus decreasing this pressure as well. For example, the use of lightweight materials in the ground behind the sheet piles can reduce the weight per unit volume, and cementation can increase the apparent adhesion. A method of cement solidification using a lightweight mixture of air bubbles can reduce the unit weight and increase the apparent cohesion; this technique can be regarded as a ground improvement method aimed at achieving both effects. As regards the passive earth pressure, a possible countermeasure considers Eq. (2.3). Ground compaction increases the angle of internal friction, which also increases the coefficient of passive earth pressure; hence, the passive earth pressure can be increased.

The bearing capacity problem must be considered in the construction of gravity-type coastal structures; checking whether the ground can support heavy loads is necessary. The bearing capacity problem is also a type of stability problem. For example, consider the ultimate bearing capacity, q, by Terzaghi's formula. The bearing capacity of a strip foundation can be expressed as follows:

$$q = cN_c + \frac{1}{2}\gamma BN_\gamma + \gamma DN_q, \tag{2.4}$$

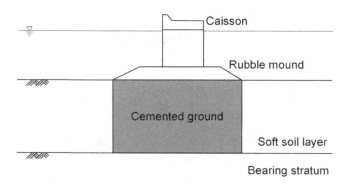

Figure 2.4 Ground improvement to increase bearing capacity.

where N_c, N_y, and N_q are the bearing capacity coefficients; c is the soil cohesion; y is the unit weight of the soil; B is the foundation width; and D is the penetration length. The bearing capacity coefficient increases with the angle of internal friction of the soil. When this angle is virtually zero (as in the case of a clay layer), N_y is also virtually zero. Moreover, when the penetration length is neglected, the ultimate bearing capacity can be simplified, as follows:

$$q = cN_c. \tag{2.5}$$

The ultimate bearing capacity is directly proportional to soil cohesion. If the bearing capacity of unimproved ground is insufficient, ground improvement measures can be implemented to increase the cohesion, as shown in Eq. (2.5) (Figure 2.4).

Piles, such as those for piers, have been widely used in coastal areas. In the past, piles were generally presumed to merely transfer loads to a solid supporting layer because the upper soil layer could be soft. However, in recent years, piles are required to share in resisting the horizontal force caused by seismic forces and the traction force of ships. If the force acting on the piles in soft soil is extremely large, the piles may break. Therefore, the lateral resistance of the upper soil layer is typically required. A method to reduce the bending moment generated in the pile as a combined pile structure is available; however, if the ground is extremely soft, the application of ground improvement must be considered.

2.3.2 Settlement and Displacement

In addition to stability problems, another critical problem associated with soft ground is consolidation settlement. In certain cases, the settlement of reclaimed land exceeded several meters. The water content of young alluvial clays is typically high, for example, approximately 100%. This means that the mass of water in the soil and mass of soil particles are nearly the same. When clay is subjected to an overburden load, the water is gradually pushed out, and the soil volume is reduced, resulting in consolidation. The amount of consolidation, S, of the clay layer is proportional to the load, p, and clay layer thickness, H. Hence, it is expressed as follows:

$$S = m_v pH, \tag{2.6}$$

where m_v is the parameter (called coefficient of volume) obtained from consolidation tests; its value is small for sandy soils but high for soft clayey soils. The problem with consolidation settlement is not the amount of settlement in particular but the number of years (possibly a decade or more) that elapses for the soil to fully settle. If the settlement is instantaneous during the construction period, filling the area with soil equivalent to the amount of settlement is sufficient. However, the occurrence of settlement after the structure has been built can have a substantial detrimental effect on the superstructure. Consolidation occurs over a long time because of the low hydraulic conductivity of cohesive soils; this means that the drainage of water from the soil occurs over an extended time. The ratio of settlement at a given time to the final settlement is called the degree of consolidation, U. It is a function of the time coefficient, T_v, and defined as

$$T_v = c_v t / H^2, \tag{2.7}$$

where t is time, and H is the thickness of the clay layer (or half of it). The coefficient of consolidation, c_v, can be expressed by the following equation:

$$c_v = k / m_v \gamma_w, \tag{2.8}$$

where k is the hydraulic conductivity, m_v is the coefficient of volume compressibility, and γ_w is the unit weight of water. The coefficient of consolidation is large for sandy soils and small for clayey soils; the smaller the value of c_v, the longer the time required for consolidation, as shown in Eq. (2.7). Based on Eqs. (2.7) and (2.8), the amount of the final settlement of sandy soil is small, and the consolidation time is brief; the opposite is true for clayey soils. In particular, the effect of the clay layer thickness, H, is significant. For instance, when the layer thickness exceeds 10 m, consolidation frequently spans over a period of ten years. In addition, a component of consolidation, called secondary consolidation, which does not follow Terzaghi's theory, has been observed to continue indefinitely. Consolidation settlement is not only caused by overburden but also by the lowering of the groundwater table. In this case, consolidation settlement is called subsidence because of its wide extent; consequently, the degree of damage is enormous. Consolidation is

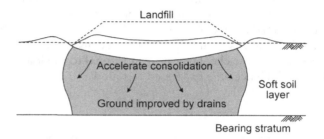

Figure 2.5 Ground improvement to prevent consolidation and settlement after commencement of operation.

not uniform from place to place, resulting in uneven settlements. Damage to structures is typically more severe when uneven settlement occurs compared with that in the case of uniform settlement. Furthermore, when a pile is driven into a solid bearing layer, and the soft clay layer above consolidates, the clay layer around the pile hangs on the pile, causing negative friction. In certain cases, this negative friction may exceed the load supported by the pile, causing the pile to bend and the superstructure to fail.

When consolidation settlement is a problem in design, the application of ground improvement, as shown in Figure 2.5, is considered. The most common ground improvement method used in the past is the drainage method, which reduces the apparent thickness of the clay layer, H, shown in Eq. (2.7), thus reducing the drainage time. In the drainage method, a material with acceptable drainage properties (e.g., sand, paper, and fibre) is inserted vertically into the ground from the surface such that water is only drained from the material and not from the top and bottom of the clay layer. Thus, the apparent thickness of the clay layer is reduced, and the time required for consolidation is shortened. Other methods for coping with the problem of consolidation are also available. Such methods include the deep mixing method, which can strengthen the ground, and the sand compaction pile (SCP) method, which accelerates drainage. In these methods, the time required for consolidation can be shortened by decreasing m_v in Eq. (2.8) and increasing c_v; in the case of sand compaction pile, k must be increased.

When soft ground is subjected to eccentric tilting loads (such as embankment loading), vertical consolidation settlement, settlement, and lateral displacement due to shear deformation can occur. In the case of large seismic forces, the soil undergoes shear deformation, plastic strain accumulates during vibration, and deformation remains after the vibration. The amount of deformation must also be considered in the design. In the past, the quantitative prediction of ground deformation was difficult. Nowadays, numerical analysis techniques based on the finite element method (FEM) have rapidly developed, enabling the evaluation of deformation through numerical simulation. The FEM is a numerical method utilised to obtain approximate solutions of differential equations that are difficult to solve analytically. The domain to be calculated is divided into small elements, and the equations in the elements are approximated by relatively simple and common interpolation functions to obtain solutions for the differential equations. This method can be used to solve not only the shear deformation problems but also the consolidation settlement, stability, and liquefaction of soil. This method is expected to be more widely used in the future. Shear deformation can be reduced by cementing the ground or using compaction techniques to increase soil stiffness. After the implementation of ground improvement, predicting the deformation of the structure using numerical simulation is preferred.

2.3.3 Liquefaction

In earthquake-prone countries, ground liquefaction must be considered in the design. In loose sandy soils, the skeletal structure of the soil is unstable. The skeleton collapses and shrinks in volume due to shear stress and impact as it attempts to form a stable structure; this is known as negative dilatancy. When the sandy soil is saturated, the volume cannot be reduced, and excessive pore water pressure is generated. When the pore water pressure is equal to the soil cover pressure, the force acting among the soil particles becomes zero, and the bonds among particles are broken, resulting in a liquid-like state. When liquefaction occurs, light structures (such as buried pipes) are lifted up, whereas structures with a specific gravity heavier than that of mud sink. Even if liquefaction does not occur, as the pore water pressure increases, the effective stress decreases. Consequently, the strength of the sandy soil decreases, leading to stability, bearing capacity, and deformation problems.

When the sandy soil is unsaturated, liquefaction is less probable to occur because the pore water pressure does not increase significantly; instead, the soil volume shrinks, resulting in the so-called 'shaking-in' settlement. In the case of saturated sandy soil, the soil settles after the pore water pressure increases and dissipates, resulting in settlement below the original level. In addition, liquefaction often causes sand boiling, leading to further ground subsidence; evidently, liquefaction is associated with subsidence problems.

To investigate whether liquefaction can occur in unimproved ground, first, the soil grain size must be compared with the particle size range of liquefaction-prone soil (0.1–1.0 mm). If the soil grain size is within this range, the safety factor associated with the occurrence of liquefaction is determined. There are various methods for determining this factor; however, the basic principle is to use the ratio of the in situ soil liquefaction resistance to the potential shear stress. For example, in the Japanese design codes for coastal structures, liquefaction is determined in two steps. First,

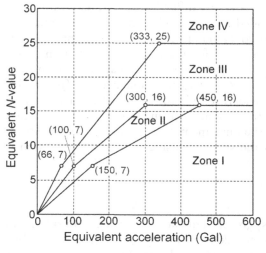

Figure 2.6 Empirical chart to determine liquefaction. (Adapted from Ref. [1].)

the resisting force is replaced by the N value from the SPT, and the shear force is replaced by the response acceleration from the seismic response analysis. Then, these values are applied to an empirical diagram to determine the possible occurrence of liquefaction (Figure 2.6). In this diagram, the smaller the N value and the higher the response acceleration, the higher the probability of liquefaction occurrence. If the assessment is ambiguous or a more detailed investigation is necessary, the degree of soil liquefaction susceptibility (liquefaction strength) is investigated by cyclic triaxial tests under undrained conditions using soil samples collected in situ. The ratio of liquefaction strength to the shear stress predicted from the seismic response analysis is then calculated for comparison to determine whether the soil can liquefy.

If liquefaction is expected to occur, countermeasures against liquefaction are considered. Some countermeasures include allowing liquefaction while strengthening the structure instead or installing piles to prevent the flow of liquefied ground. In recent years, however, ground improvement methods, such as sand compaction pile, have been widely used to prevent liquefaction. Compaction makes the soil denser and more resistant to liquefaction by increasing the confining pressure. As the N value increases with compaction, this value must be reapplied to the empirical diagram shown in Figure 2.6 to design the soil such that it does not liquefy. The deep mixing method is also employed to reduce the shear stress in soil during an earthquake by surrounding the liquefiable ground with a grid [2, 3]. The narrower the spacing among the grids, the less the probability of liquefaction occurrence.

The narrower the spacing, the higher the rate of improvement, and the less economical the method. Accordingly, the appropriate grid spacing to reduce liquefaction must be proposed, and the grid spacing to be implemented must be set according to this proposal.

2.3.4 Problems Caused by Water

Liquefaction is not the only problem associated with loose sandy soils. Seepage failure due to considerable differences in water levels in front of and behind the embankment can occur, or wave lift pressure may be experienced. Boiling and piping are also critical considerations in the design of temporary earth-retaining sheet piles. These phenomena can be regarded as the replacement of excess pore water pressure by seepage water pressure in liquefaction; the principles and phenomena are similar to some extent. For example, when sheet piles are driven into soft ground, the equation of boiling by Terzaghi is expressed as follows:

$$2\gamma d / \gamma_w h_w > 1.2 - 1.5, \tag{2.9}$$

where γ is the unit weight of soil in water; d is the penetration length; γ_w is the unit weight of water; and h_w is the water level difference. The value on the right-hand side of the equation varies according to the design criteria of each structure. However, its ratio is generally considered to be 1.2–1.5. The foregoing equation indicates that the safety against boiling can be increased by increasing the length, d, of the embedment and decreasing the water level difference, h_w. If boiling is expected to occur, increasing the penetration length of sheet piles or reducing the water level difference (using deep well or well point methods) is conventionally applied.

In recent years, ground improvement methods have been widely used in this problem. Among these is the jet grouting method using a high-pressure jetting agitator, which allows the soil in the immediate vicinity of the structure to be consolidated (Figure 2.7). This technique leverages the adhesion between the structure and solidified soil to resist the water pressure penetration from below and prevent boiling. The foregoing is equivalent to increasing the numerator on the left-hand side of Eq. (2.9), which represents the resisting force. Dry space is formed because the solidified soil blocks water flow. A wider application of this method is anticipated in the future.

2.4 EXAMPLES OF DESIGN PROCEDURE

The previous section provided an overview of the design methods, items to be considered in the selection of improvement methods, and aspects that must be accounted for in the design. However, fully comprehending the design process without examining specific procedures can be difficult.

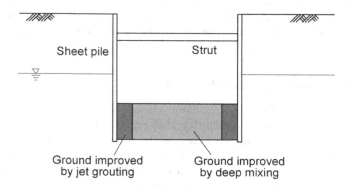

Figure 2.7 Jet grouting method in vicinity of structures.

Introducing the design procedure for a gravity-type quay wall (one of the typical structures in coastal areas where the original ground is to be improved) as a reference may be advantageous. This section presents the design procedure for quay walls. The actual design process must be referred to appropriate design standards pertaining to the conditions of a coastal structure. This is because design procedure details vary according to the type of structure and country in which construction is implemented.

In Chapter 1, the overall project flow in the construction of a quay wall to be constructed at a specific location based on a port development plan is shown in Figure 1.6. Two steps are involved: design of the original ground and design implementation after determining the ground improvement method. Figure 2.8 shows the details of each step. First, a quay wall cross section is assumed based on the design conditions. Subsequently, the stability of the wall is verified without considering the behaviour of the underlying ground by determining whether the wall could slide or overturn and the foundation pressure exceeds the bearing capacity. Moreover, the wall parameters are determined such that the foregoing does not occur. Once the wall specifications have been determined, the next step is to design ground improvement. Section 2.3 describes the soft ground problems ((1) stability, (2) deformation, (3) liquefaction, and (4) water-related problems), particularly problems 1–3, which affect gravity-type quay walls. The overall stability of the quay wall, including the ground, is verified by assuming that ground improvement is not implemented; if it was, ground improvement parameters were assumed. The calculations were repeated while reviewing the ground improvement parameters to ensure that stability requirements are satisfied. If realistic conditions for ground improvement cannot be found in terms of economy and workability, reconsidering the assumptions regarding the quay wall cross section in the first step is necessary. If ground

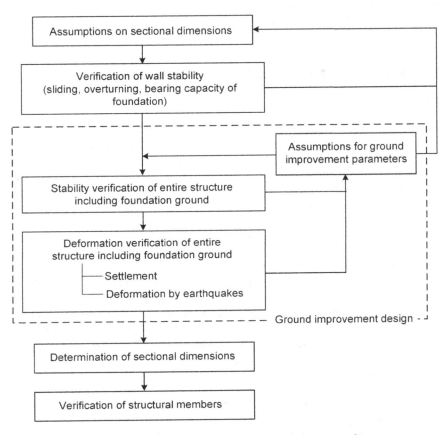

Figure 2.8 Design procedure of gravity-type quay walls including ground improvement.

improvement parameters satisfy the stability requirements, the next step is deformation verification; the same process described above applies. The calculations are repeated while reviewing the ground improvement parameters to ensure that performance is satisfactory. Deformation verification typically considers two items: structure settlement and quay wall displacement during earthquakes. Even if the stability is satisfied, the use of a quay wall is difficult if it settles significantly or moves seaward during an earthquake; hence, deformation verification is an important step. Ground softening due to liquefaction during earthquakes is frequently considered. For gravity-type quay walls, the verification of water-related problems is usually ignored because the occurrence of these problems is improbable. However, in the case of seawalls and sheet pile quay walls, these problems may be included in the verification. Accordingly, the quay wall cross section including ground improvement is considered by verifying the stability and deformation of the entire structure in addition to the stability of the wall itself.

2.5 PHYSICAL AND NUMERICAL MODELLING

2.5.1 Physical Modelling

Structural or ground improvement design can be implemented in accordance with existing design procedures provided that the experience in the construction of a coastal structure type is considerable or the characteristics of the ground and the applied external forces are well defined. In contrast, if a different structure is to be constructed or a new ground improvement method is to be applied because of social requirements, such as economic or environmental reasons, model tests may be employed to confirm the performance of the structure or ground improvement. Although model tests are often considered for research and technological development, they can also be used in design. Model testing and numerical analyses are typically combined for design purposes. In this approach, a phenomenon whose mechanism is not understood is reproduced through a model test, and the methods and conditions of the numerical analyses are adjusted to reproduce the behaviour. Then, the behaviour of the real structure is simulated by numerical analysis. Because reproducing the real structure in a model test is difficult, numerical analysis is used as a medium. Specific examples are discussed below.

Simplification and scale reduction may be the challenges of model testing for reproducing the behaviour of real structures. Real ground is complex and virtually impossible to reproduce in a model. Simplification is achieved by omitting secondary factors affecting the ground behaviour, leaving direct factors in place to highlight the behaviour to be reproduced. For example, if determining the effect of a vertical drain is desired, reproducing a complex multi-layered ground is unnecessary; a model with drain material in low-permeability ground can be sufficient. Problems related to size reduction are also encountered. Because a small model reproduces the behaviour of a prototype-scale structure, determining the ratio (similarity ratio) between the prototype and model in terms of physical quantities, such as stiffness, strength, weight, and time, in addition to dimensions is necessary. The problem is that model tests in a gravitational field cannot reproduce ground stress, which has a significant influence on ground behaviour. Moreover, assumptions must be included or partially relaxed in the similarity rule. Assumptions are also made regarding the behaviour of the prototype structure predicted by the model test. This problem can be resolved using the centrifuge model test method in which testing is conducted in a centrifugal force field. By applying centrifugal forces to the model, producing the same stress and stress–strain relationship as those of the prototype-scale ground is possible. Many examples have confirmed that the results of centrifuge model tests are consistent with those of the real ground. Although this test has some disadvantages, its advantages outweigh these flaws.

A case study involving the author is presented as an example of the use of model tests in design. A breakwater was planned to be constructed in a bay to prevent the entry of tsunamis and rise of sea water level. After various studies, the decision was to adopt a gravity-type breakwater. Based on a ground investigation, the underlying ground was assumed to liquefy during a large earthquake. However, if the ground liquefied, the breakwater could settle significantly, and the risk was that the required top height could not be secured against a tsunami that could strike after an earthquake. The settlement of the breakwater was difficult to prevent because of ground and local conditions. Although the entire liquefied layer requires improvement, budgetary constraints must be considered. Consequently, the decision was to only partially improve the shallow section. However, predicting the extent of the settlement of the breakwater due to the ground liquefaction caused by an earthquake was difficult. Accordingly, centrifuge model tests were conducted to simulate the ground liquefaction and breakwater settlement caused by an earthquake. The centrifuge model test apparatus [4] and cross-sectional view of the model are shown in Figure 2.9. A shaking table and model were placed on the platform of the apparatus, and vibrations were induced when centrifugal forces were applied to the model to liquefy the ground and apply seismic forces to the breakwater. Figure 2.10 shows the breakwater before and after double shaking. In this test, the settlement of the breakwater was 1.0 m on a prototype scale. Although the dimensional ratios are the same as those of the prototype structure, whether the actual settlement can be adequately predicted cannot be ascertained because the soil material in the test differed from the real one. The results of the model tests were reproduced using finite element analysis (FEA), which can also be used for dynamic analysis. Various parameters were adjusted such that the degree of liquefaction and amount of settlement matched. The settlement was then estimated via numerical analysis using the soil properties in the field. In the design, the height of the breakwater was determined such that the prescribed top height could be maintained even if settlement occurred. Without the model tests, the design would have relied on the results of the unreliable numerical analysis, which would have been extremely costly to ensure safe design, or the design may not have been possible in the first place. Thus, model tests can be effectively used in design.

2.5.2 Numerical Modelling

Ground behaviour is complex and typically cannot be reproduced by simplified models. Many reliable formulae have been proposed for static stability

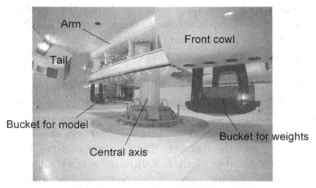

(a) Centrifuge machine (PARI Mark II-R)

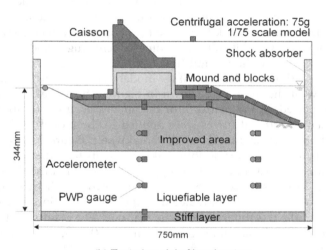

(b) Tested model of breakwaters

Figure 2.9 Centrifuge machine and schematic of model.

problems, such as bearing capacity and earth pressure. Research has also been conducted on one-dimensional consolidation settlements, enabling the estimation of the amount of settlement possible to some extent. However, the estimate of two-dimensional (2D) deformation, dynamic ground behaviour, and liquefaction problems using simplified models is generally difficult. Thus, to estimate deformation, numerical analysis is employed. Many numerical analysis methods have been formulated including finite difference method, FEM, boundary element method, finite volume method, and distinct element method. New types of analysis using the particle method

(a) Before shaking　　　　　　　　(b) After shaking

Figure 2.10 Breakwater settlement due to seismic motions.

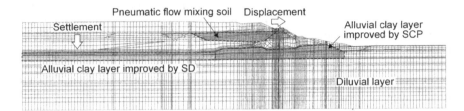

Figure 2.11 Seawall deformation calculated by FEA.

are smoothed-particle hydrodynamics and moving particle semi-implicit. Among the techniques, FEA is most widely used for geotechnical design at a practical level. The FEA is employed to numerically obtain approximate solutions to differential equations that are difficult to solve analytically. The area to be calculated is divided into small elements, and the equations in these elements are approximated using relatively simple and common interpolation functions to obtain solutions to the differential equations.

Figure 2.11 shows a case study of the deformation that occurs during the construction of a seawall using FEA. The example includes the results of the deformation analysis of a seawall on reclaimed land on which the fourth runway (Runway-D) of Tokyo International Airport is constructed. The long-term deformation of the seawall after construction and reclamation was investigated. Information on 2D deformation, such as the downward vertical displacement and horizontal displacement to the sea caused by the shear deformation of the soil, was obtained by FEA. In addition, FEA outputs sequential deformation, stress state of soil, and excess pore water pressure, rendering the causes of deformation easy to explore. For example, a method to reduce plastic shear strain in an alluvial clay layer must be considered in design if this strain can cause the horizontal displacement of the seawall. Methods for reducing the plastic shear strain include increasing the soil strength using deep mixing or the SCP method. At this site, building an embankment with rubble stones as seawall on the original ground and reclaiming the area behind the embankment were planned. However, the preliminary design studies clearly indicated that the original ground was soft. Moreover, if the embankment was constructed as designed, problems involving bearing capacity, settlement, and deformation could occur. Thus, the soft ground below the embankment was improved using the SCP method. To increase the overall stability of the embankment and reduce the amount of deformation, the soil behind the embankment was treated using a pneumatic flow mixing method.

Numerical analysis is frequently used to investigate the seismic behaviour of coastal structures. Predicting the failure and deformation of structures caused by earthquakes is difficult although the static stability of structures can typically be assessed using simplified models. In particular, formulating and determining the deformation of the structure including the degree of liquefaction are difficult if the ground is liquefied. Numerical analysis using FEA is typically used for this purpose. A FEA program called FLIP [5, 6], which is a dynamic effective stress analysis program that considers liquefaction, is generally used in the design of coastal structures in Japan. Quay walls that can be quickly restored and put back into service after a major earthquake are called earthquake-resistant quay walls. The design standard stipulates that the displacement of quay wall must be limited to approximately 1 m. At the design stage, the displacement was confirmed to be within this limit using FEA. An example of the calculation is shown in Figure 2.12, which shows the deformation diagram after the seismic input. If the displacement is large, the application of ground improvement is considered to avoid liquefaction, and its effect is verified through numerical analysis. The extent of improvement was also determined by parametrically varying the upgrade conditions.

Numerical analysis is similar to a 'black box'; once initial and boundary conditions have been entered, ground deformation is automatically

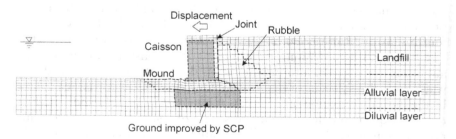

Figure 2.12 Deformation of quay walls calculated by FEA. (Courtesy of Japan Port Consultants.)

outputted. The calculation results are graphically attractive and can be easily attributed to the plausible deformation of the ground. The use of proven numerical analyses is important for the design; the analysis results have been compared with those of actual disaster cases and model tests. Numerical analysis is complex and allows the input of many parameters. It is typically considered an excellent method because of its easy fitting to and satisfactory consistency with test results. However, it does not consistently yield superior results in predicting the movement of actual structures because it makes many assumptions when predicting the behaviour of structures. Nevertheless, numerical analyses, which have long been used and yielded satisfactory and reliable results, must be applied to design.

2.6 EXECUTION

2.6.1 Use of Information Technology

After planning and design, the next step is on-site execution. Because the execution of ground improvement differs according to the various execution methods, it is introduced starting from Chapter 3 onwards in each section. In this chapter, two common points in execution are explained: effective use of information technology and execution management. The introduction and use of information technology have progressed in the field of ground improvement. The introduction of information and communication technology (ICT) and the automation of execution can be cited as representative examples of this advancement. Information technologies are actively introduced into execution machines on land with the use of ICT to display information during execution (such as the depth of the agitator blade, penetration and withdrawal speeds, rotational speed, and cement slurry flow rate) in animated form. They allow the constant monitoring of information using the tablet terminals located outside the execution machine. The use of augmented reality (AR) technology is also enabled to display information regarding buried objects and the execution location on the screen (Figure 2.13). A maritime execution machine has also been introduced to the technologies. For example, in the SCP method, the practical application of technology has been implemented to assist the machine operator by obtaining information on the position of the sand pile centre from the global navigation satellite system. Information technology that supports work and improves the certainty of execution is expected to be introduced in the future. Furthermore, the automation of land machine execution is possible. The introduction of information technology to execution not only improves productivity but also reduces the variations in execution accuracy owing to differences among operators. However, overdependence on information technologies is dangerous. Unexpected errors in the programming

Figure 2.13 Example of using AR technology. (Courtesy of Fudo Tetra Co.)

execution procedure may be encountered; hence, allowing the machine to learn through repeated trial and error is important. Further, converting execution results into big data and allowing the machine to learn using artificial intelligence technology are critical.

Other measures for enabling the use of information technology include the visualisation of ground information and improvement status. As part of the application of building information modelling (BIM), which has been developed in the construction engineering field, a 3D model integrating the project management of ground improvement works has also been utilised. This has enabled the sharing of information among concerned parties, thereby increasing the efficiency and sophistication of ground improvement works. The technological development in this respect has also advanced with respect to land works. For example, the display of information after ground improvement (e.g., improvement depth, improvement position, motor current values, and cement slurry flow rates) in a 3D model is currently possible, allowing the visual confirmation of how the work has been implemented. The incorporation of survey information regarding actual ground improvements is also important. This is because the execution record and actual improved ground may differ. For example, attempts have been made to properly understand the shape of improved ground by geophysical exploration.

The use of information technology is rapidly advancing, and the description provided herein may soon become obsolete. Hence, to make appropriate decisions on whether the information presented can be used in practice and to select as well as use it accordingly, keeping abreast of the latest information is important.

2.6.2 Execution Management

As mentioned, various investigations are required in the planning and design stages; however, investigations are also necessary in the execution stage. Here, investigations refer to execution management; that is, the measurement management of the execution (also known as quality control) and quality assurance of the improved ground. Measurement management is explained as follows. From the start to the end of implementing ground improvement (or when the effect of ground improvement is achieved), determining the behaviour of the ground and structures as well as the effect on surrounding ground and coastal structures is necessary. Based on the measurement results, ascertaining whether the behaviour of the ground and structures is what is anticipated according to the design and ground improvement is required. In particular, the fact that ground improvement is implemented indicates that the ground on which the structure is to be built is soft. Comprehending the behaviour of the improved ground, its structure, and its effects on the surrounding area is important. Design is only a desk study, and the anticipated behaviour typically differs from the real one; hence, reviewing the design once in the execution process is necessary. For example, a modified design may be implemented, or the execution plan may be revised if necessary, as shown in Figure 2.14. Note that this cycle may be repeated to ensure that suitable coastal structures are constructed.

An important but difficult part of measurement management is determining measurement items, points, and periods. Table 2.3 summarises the specific measurement items. The implementation of all the listed items is not necessary; however, the selection of appropriate items from the list depending on the type of coastal structure and ground improvement method is required. In selecting the measurement items, first, clarifying the purpose of the management is necessary. In addition to controlling the execution, management may be implemented for ground improvement quality control or conducting research. In most cases, ground deformation is measured, and pore water pressure may be determined to ascertain consolidation. Groundwater levels may be measured to check for obstructions

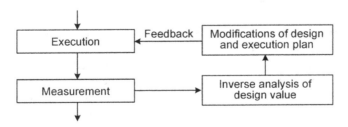

Figure 2.14 Design and planning review flow during execution.

Table 2.3 Measurement items during execution

Item	Purpose	Method
Change in soil properties	• Confirm water content and strength increased by consolidation	• Sampling • In situ test
Vertical settlement	• Contrast with design settlement and estimate final settlement • Sense sign of ground failure	• Subsidence meter • Stratified settlement meter
Horizontal settlement	• Check shear deformation • Sense sign of ground failure	• Surface displacement pile • Surface extensometer • Insertion inclinometer
Pore water pressure	• Confirm drainage properties • Estimate consolidation speed • Check applied load	• Pore water pressure gauge • Water level measurement
Earth pressure	• Check working load	• Earth pressure gauge
Others	• Check displacement of surrounding area • Check water level and quality of groundwater	• Settlement meter and inclinometer • Water level measurement and water quality test

to groundwater flow. The candidate measurement points may include those important for the performance of the structure; representative points where the stability of the ground can be assessed based on the failure modes considered in the design; and points where the impact on the surrounding area can be assessed. Regarding the measurement period, the continuation of measurements during execution is important. The measurement intervals must be close when external forces are applied (for example during ground improvement, reclamation, construction of embankments, or installation of structures).

Quality assurance of ground improvement is another important aspect. The success or failure of ground improvement considerably depends on the execution technique. If the execution is inadequately performed, the assumptions are broken, and the effects of ground improvement are not demonstrated regardless of how cautious the investigation and design are implemented. Thus, ensuring the appropriate quality and analysing as well as evaluating the results of the execution for the next construction project are necessary. Ground improvement involves two quality assurance items: the specified shape is achieved, and the targeted strength and other properties are obtained. In the ground improvement for coastal structures, visually checking whether the improved ground is formed according to a specified shape is difficult because it is typically underwater. For this reason, quality is usually ensured from the data measured during the execution and from records. By combining the 3D information from geophysical surveys and line information from borings, confirming the shape of the improved ground is also possible. Targeted characteristics, such as strength, can be checked by implementing ground borings and soundings, as described in the section on investigation. The results directly indicate whether the prescribed strength has been achieved and variation is within an acceptable range.

Execution management, such as measurement management and quality assurance of improved ground, have been described. Moreover, the measurement of some improved ground and coastal structures must be continued after execution. This is to ensure that the required functions of the improved ground are appropriate and effective as well as maintained on an ongoing basis. For example, if reclamation is to be implemented on soft ground, the vertical drain method is used to accelerate consolidation and protect the surrounding ground. Although most of the consolidation settlement is assumed to have been completed by the time the improved ground is put into service, allowing some time for settlement to continue thereafter is a common practice because the completion of consolidation is overly long. In this case, the continuous measurement of the settlement is necessary to ensure that the structure is within the specified allowable settlement.

References

1. Ports and Harbours Bureau, Ministry of Land, Infrastructure, Transport and Tourism (MLIT). 2018. *Technical Standards and Commentaries for Port and Harbour Facilities*. The Ports & Harbours Association of Japan (in Japanese).
2. Babasaki, R., Suzuki, K., Saitoh, S., Suzuki, Y., and Tokitoh, K. 1991. *Construction and Testing of Deep Foundation Improvement Using the Deep Cement Mixing Method. Deep Foundation Improvements: Design, Construction, and Testing*. ASTM International.
3. Takahashi, H., Kitazume, M., and Ishibashi, S. 2006. Effect of deep mixing wall spacing on liquefaction mitigation. Proceedings of the 6th International Conference on Physical Modelling in Geotechnics, 585–590.
4. Takahashi, H., Fujii, N., Morikawa, Y., and Takano, D. 2019. Development of hydro-geotechnical centrifuge PARI Mark II-R. *Technical Note of the Port and Airport Research Institute*, No. 1353, 27p.
5. Iai, S., Matsunaga, Y., and Kameoka, T. 1992. Strain space plasticity model for cyclic mobility. *Soils and Foundations* 32(2): 1–15.
6. Towhata, I. and Ishihara, K. 1985. Modeling soil behavior under principal stress axes rotation. Proceedings of 5th International Conference on Numerical Method in Geomechanics, 523–530.

Chapter 3

Ground Improvement Methods

In this chapter, various in situ ground improvement methods for soft ground in coastal areas are presented. Soft ground improvement has been originally intended to resolve soil stability and deformation problems during construction. Nowadays, soft ground improvement is increasingly implemented to enable the subsequent construction of structures using strengthened soil. Ground improvement for coastal areas is discussed in this chapter, and construction after implementing ground improvement is presented in the subsequent chapter. The ground improvement methods introduced in this chapter are soil replacement, consolidation, and compaction. Cement mixing and chemical grouting for soil solidification are also discussed. Each method is initially presented with an overview; then, investigation, design, and execution are elucidated.

3.1 REPLACEMENT METHOD

3.1.1 Overview

The problems related to a soft ground have been described in the previous chapters, and the most readily conceivable ground improvement method is to replace the soft ground with soil that will not cause problems. The replacement method involves the partial or complete removal of the soft ground and its replacement with high-quality soil, and this method has been used in both sea and land areas as it can be performed quickly and reliably. Replacement methods can be divided into two categories: excavation and forced replacements. When excavation replacement is conducted in a coastal area, soft soils are excavated and removed by pumps or grab dredgers and backfilled with high-quality soil. In contrast, in forced replacement, the soft soil is forced out laterally by the weight of the embankment or via blasting and replaced with high-quality soil. The former method is predominantly used in coastal areas because of the difficulty in estimating the extent of improvement in the latter method. However, in recent

DOI: 10.1201/9781003267119-3

years, the use of replacement methods in coastal constructions has been decreasing because it causes turbidity in the sea; the disposal of excavated soil is challenging and, primarily, there is a worldwide depletion of high-quality soil such as sand. Moreover, soft clay soil is generally replaced by sand; sand dropped into the water is loosely deposited, which may result in insufficient support for large structures and liquefaction damage in countries where earthquakes occur. If recycled materials or artificial sand can be used as replacement materials, these methods may once again attract attention; currently, the number of cases in which these techniques are used is decreasing.

3.1.2 Investigation and Design

Geotechnical investigations required for the design of replacement methods include geophysical, consolidation, and shear tests of the sampled soil. The physical properties and shear strength of the soil are used to determine the excavation method and slope after excavation. Consolidation tests are especially performed when the ground below the excavated soil is a clayey layer and are used to estimate the consolidation settlement after replacement. The physical properties of the excavated soil should also be verified to ensure compliance with the recipient standards. The depth and stratification of the soft ground should also be confirmed by soundings and borings. These data provide important information for determining the replacement depth.

For the design, design conditions and cross-sections (replacement depth, replacement width, and excavation gradient) are assumed, and studies on slip failure, settlement, and selection of replacement material have been conducted. Consideration of the liquefaction potential of the replacement material and an assessment of its impact on the superstructure may also be required. The replacement depth should be set as follows: if the soft layer is relatively thin, the entire layer should be replaced; if the soft layer is thick, the depth at which the subgrade vertical stress due to loading is experienced should be less than the bearing capacity of the ground, and this constraint is used as a guide for replacement. To determine the replacement depth, the feasibility from the perspective of the capacity of the execution machinery should also be verified. Then, the replacement width should be determined. The length of the short side of the trapezoid (replacement width) can be determined by referring to previous execution examples, as shown in Figure 3.1. The excavation slope is determined by considering the strength of the original ground and excavation depth, which, according to previous examples, is generally 1:1.5–3.0.

Once the extent of the improvement can be assumed, the stability against external loads should be examined using calculations such as circular slip analysis. When sheet piles or anvils are installed in the replacement

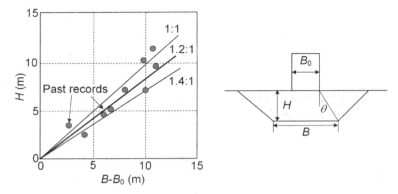

Figure 3.1 Relationship between replacement width and depth. (Adapted from Ref. [1].)

section, in addition to the normal earth pressure calculations, a composite slip study should be conducted to calculate the earth pressure. When the bottom of the excavation is sloped in a full-layer replacement, a composite slip study including sliding failure at the bottom of the excavation should be conducted. Although the requirements for replacement materials are not clearly defined, a wide grain size range and low fine fraction content below silt content are desirable. In previous execution cases, mountain sand was used in some cases where the fine fraction content was more than 20% by weight; however, in general, the fine fraction content is often set at 15% or less. The internal friction angle of replacement materials is generally considered to be approximately 30°; this value is affected by the grain size and distribution of the replacement material, feeding method, feeding sequence, leaving period, and loaded weight, and may be low in some cases. When N values were investigated using the standard penetration test (SPT), they were approximately 10 for the instantaneous feeding of a large volume of sand using a large-capacity barge, approximately 5 for feeding using a gutted barge, and even smaller values for feeding using a pump barge. In certain cases, N values increased when a load was applied or when the period of abandonment was prolonged.

If clay remains below the replacement section (partial replacement or below the excavation slope), consolidation settlement is expected to occur in the clayey section; therefore, the settlement should be estimated by considering the consolidation yield stress of the clay layer and the vertical load acting on it, and whether the settlement will have any adverse effect on the superstructure. To estimate the amount of settlement, the vertical load dispersion at the bottom of the replacement section should be considered, and the m_v, C_c, and e–log p methods, which can be found in the textbooks on soil mechanics, should be used.

3.1.3 Execution

The replacement method used in coastal areas is generally excavation replacement, where the soft soil is first excavated with a pump or grab dredger. During excavation, the ground is softer than expected, and the slope gradient may not be maintained as designed; therefore, excavation is performed while measuring the displacement of the surrounding ground to ensure that the slope does not collapse. After excavation, a ground investigation should be conducted to verify whether the strength of the excavation bedding surface satisfies the design values. Barges, gutters, and pumping vessels are then used to pump in high-quality soil, and a depth survey is performed to confirm the final volume. When the soil is placed underwater, it is unlikely that the ground is loosely deposited. As a quality control, ground investigation should be conducted after execution. Compaction after placing the replacement materials would require the use of another ground improvement method; however, this is impractical because of the high costs involved. If ground strength cannot be expected after placing sand, advance measures must be considered, such as mixing cement with the soil to be placed.

3.2 SURCHARGE AND DRAINAGE METHODS

3.2.1 Vertical Drainage

3.2.1.1 Overview

The problems with soft cohesive soils are their low strength, which makes it impossible to ensure the stability of structures on the ground, and the long-term displacement of structures owing to consolidation settlement and two- or three-dimensional ground deformation. The reason for the low stiffness and strength of clayey soils is that they contain a large amount of water; the stiffness and strength can increase if the soil is consolidated. However, consolidation occurs over a long period of time, ranging from several years to several decades. As it is necessary to improve the ground for rapid consolidation, a typical ground improvement method that promotes consolidation is the vertical drain method. In this method, several artificial vertical drains are placed in the clayey layer to promote consolidation by reducing the ratio of the vertical to horizontal drainage distances. Because vertical drains are cast, this method is called the vertical drain method. According to one-dimensional consolidation theory, the following equation is established between the time t required for consolidation of the clay layer and the drainage distance H' ($H'=H/2$ for double-sided drainage and $H'=H$ for single-sided drainage, where H is the thickness of the clay layer).

$$t = \frac{H'^2}{c_v} T_v \qquad (3.1)$$

From the above equation, the consolidation time can be significantly reduced by considering the thickness between the drains as H instead of the thickness of the clay layer in the vertical direction in the field as H. Since the theoretical consolidation time estimation formula for this method was proposed by Barron [2] in 1948 and the design method was established, it has been widely used worldwide and is now the most standard method of accelerated consolidation. Note that the consolidation of the ground does not proceed simply by placing a drain; therefore, it is necessary to add an overburden load by filling over the improved ground or reduce the water pressure in the drain, as described in the subsequent section.

Vertical drainage methods can be divided into two main types: sand drains (SDs) (sand pile diameter 40–50 cm) using sand as the permeable material and prefabricated vertical drains (PVDs) (drain material thickness 0.3–0.4 cm, width 10 cm) using man-made materials such as specially processed plastic material as the permeable material. When plastic material is used in the PVD-based method, the method is also called plastic board drain (PBD). In both cases, the size of the drain material (diameter and width) and the spacing between drains are important parameters in the design process because they determine the consolidation enhancement effect. The spacing between drains ranges from 1.5–4.5 m for SDs, but 2.0–2.5 m is the most common range, accounting for approximately 70% of the total distance. For PVDs, the range is 0.5–2.5 m, but a range of 1.0–1.5 m is common, accounting for approximately 80% of the total distance. Figure 3.2 shows the schematic diagrams. The SD-based method was first used in USA in the 1930s, and subsequently became common for coastal reclamation projects in Japan. In this method, a casing pipe filled with sand is inserted into the ground, and the sand is placed in the ground as the pipe is extracted. This is a highly reliable method because sand piles equivalent to the diameter of the casing pipe can be driven reliably. Moreover, the shear strength of sandy soil is higher than that of cohesive soil, and the effect of replacing cohesive soil with sandy soil is thought to increase the stiffness and strength. This effect is usually not anticipated in the design, but is a hidden effect of the SD method. The PVD method was initially developed in Europe as a method for placing paper drains that were manufactured by laminating grooved paperboards into the ground and was used to improve significantly soft landfill sites. It was subsequently observed that the permeability, which is the most important element in the consolidation promotion method, was significantly reduced owing to meandering and clogging caused by settlement, resulting in a delay in consolidation, which is a fatal defect. Subsequent improvements led to the development of the present method, which uses drains made of fibrous or

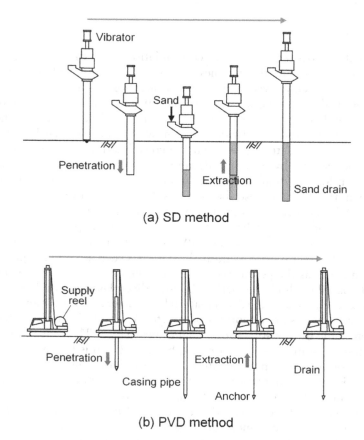

Figure 3.2 Sand drain and prefabricated vertical drain methods.

plastic materials. Today, PVDs can be used at depths greater than 50 m and this method is widely used and highly reliable.

Further, the effectiveness of SDs was heavily debated in the 1960s when large-scale ground improvements began to be implemented, because the effectiveness of SDs was less clear for land works than for reclamation works in coastal areas. The effect of SD was not as pronounced in land works as the clayey layer was not significantly thick and the stratification state was complex, with several intervening highly permeable sandy silt layers. Currently, PVD is also used; however, its effect may not be as pronounced in certain cases. In contrast, the effectiveness of vertical drainage in coastal areas is clear and is one of the most widely used ground improvement methods in coastal areas.

In addition to the vertical drain method, which is intended to promote consolidation of soft cohesive soils, the vertical drain method is sometimes used to prevent liquefaction during earthquakes. To quickly dissipate the excess pore water pressure generated in sandy soils during earthquakes, a crushed stone drain is constructed in the ground (known as the gravel drain method). Although this method should be regarded as a method to reduce liquefaction damage, because liquefaction cannot be completely controlled against large earthquakes, it has often been used for reclaimed land in coastal areas because it is an economical method for ground improvement. In this book, gravel drains for liquefaction control are not described, but they can quickly reduce pore water pressure by shortening the drainage distance, similar to vertical drains, for the accelerated consolidation of cohesive soil.

3.2.1.2 Investigation and Design

In addition to soundings and borings to determine the stratigraphic composition, geophysical, consolidation, and shear tests of the sampled soil are necessary for the design of vertical drainage. The consolidation test results are used to estimate the final settlement and consolidation times, while shear tests are used to estimate the strength. Other investigations may include tests to determine the properties of the drain material. In this test, the drain material is actually embedded in cohesive soil to check its performance. Constant head permeability tests can be conducted with compacted cohesive soil around the drain material, allowing the permeability, including the effects of loading and drain material bending, to be determined.

The design procedure is shown in Figure 3.3. The design of the vertical drainage method involved estimating the amount of consolidation settlement and time required for settlement. Once these parameters are estimated, the method can be executed while increasing the strength of the ground, for example, by staged construction in reclamation projects, and the time when the upper structure is built and put into service can be determined. In general, because natural ground is often heterogeneous, it is impossible to completely understand the ground properties to be constructed; moreover, it should be considered that progress is rarely predicted at the time of design. It is useful to conduct on-site measurements such as settlement, lateral displacement, and pore water pressure at the time of execution and compare them with the predicted values at the time of design to implement countermeasures subsequently.

The final consolidation settlement is estimated by adding the one-dimensional consolidation settlement to the settlement associated with the two- or three-dimensional deformations that occur immediately after loading. The estimation of the latter settlement is difficult using simple models and is neglected in the case of extensive reclamation. If deformations

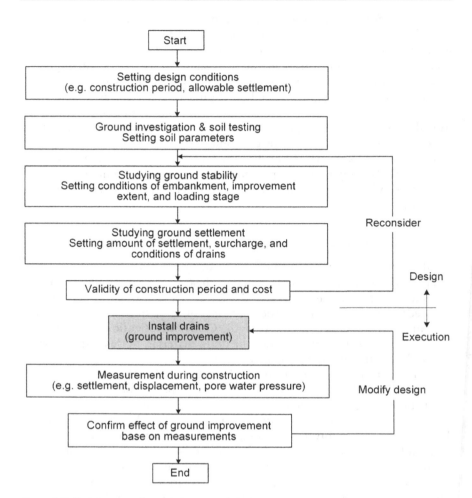

Figure 3.3 Design procedure of the vertical drain method.

cannot be ignored, a numerical analysis using the finite element method is required. For the settlement due to one-dimensional consolidation, the aforementioned simple methods such as m_v, C_c, and e–$\log p$ methods can be used for the estimation. However, the estimation of the consolidation time is challenging. If drains are installed vertically in clay, the pore water flow will be horizontal and will radiate towards the drain material, except in areas close to the drainage layer. Consolidation via the vertical drain method considers a model with a drain of diameter d_w in the centre of a clayey cylinder with diameter d_e, as shown in Figure 3.4 (r in radial coordinates, z in vertical coordinates). d_e represents the extent of clay borne by one drain

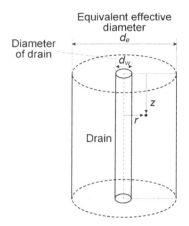

Figure 3.4 Drainage model of the vertical drain method.

and is called the equivalent effective diameter. By neglecting vertical flow, assuming infinite drain permeability, and without disturbance (smeared zone) of the cohesive soil around the drain, Barron [2] suggested that the mean degree of consolidation U can be approximately calculated by the following equations:

$$U(T_h) = 1 - \exp\left\{-\frac{8T_h}{F(n)}\right\},$$ (3.2)

$$T_h = \frac{c_h}{d_e^2}t,$$

$$F(n) = \frac{n^2}{n^2 - 1}\log_e n - \frac{3n^2 - 1}{4n^2},$$

$$n = \frac{d_e}{d_w},$$

where T_h is the time factor, t is actual time, and c_h is the horizontal coefficient of consolidation. This solution is referred to as simplified Barron's solution and is often used for vertical drainage methods. Vertical drains are typically placed in an equilateral or square arrangement. The area where pore water flows into the drain is the area of a regular hexagon or square surrounded by faces equidistant from each drain, and the diameter of a circle of equal area is the equivalent effective diameter d_e, which has the following geometric relationship with the drain casting interval d:

For equilateral triangular arrangement: $d_e = 1.05d$, (3.3a)

For square arrangement: $d_e = 1.13d$. (3.3b)

The drain diameter d_w can be used as in the case of SD. In the case of PVD, the width a and thickness b are converted from circumferential length to circular parameters according to the following formula:

$$d_w = \frac{2(a+b)}{\pi}.$$ (3.4)

To account for consolidation delays due to the disturbance of the clay around the PBD during casting, it is often empirically designed as $d_w = 5$ cm (for $a = 10$ cm, $b = 0.3$ cm). Furthermore, the horizontal consolidation coefficient c_h is often several times larger than c_v because horizontal permeability is better than vertical permeability in general soils; however, there are still some unknowns, and it is safe to assume that c_h is equivalent to c_v. Note that Barron's solution does not consider one-dimensional consolidation in the vertical direction. Therefore, when the spacing between drains is large, for example, it is necessary to superimpose Barron's solution and the one-dimensional consolidation solution to obtain the degree of consolidation. If the consolidation time can be estimated, the consolidation settlement from time to time can be obtained by multiplying the final settlement, and the shear strength at each time can also be estimated from the rate of increase in shear strength with respect to consolidation.

The design of this method should consider the permeability resistance of the drain (well resistance), stress concentration on the drain, disturbed clay (smeared zone) with reduced permeability around the drain caused by drain casting, permeability and thickness of the sand used for sand matting, and effect of these factors on the consolidation process of vertical drains. These influences are significantly complex because of the interplay of individual factors, and these concepts are discussed in technical papers. Only the well resistance and smear zones are briefly discussed here. The permeability resistance of a drain is also referred to as well resistance; moreover, a reduction in the permeability of a vertical drain causes hydraulic head loss in the drain, which affects the consolidation rate. Although it is important to deal with the drain so that it does not become clogged with soil, Yoshikuni [3] obtained a solution for the consolidation delay due to well resistance, and proposed a design method that considers the well resistance. When considering the smeared zone, drain driving is known to cause turbulence in the cohesive soil around the drain, resulting in a decrease in the horizontal permeability and consolidation coefficient. Generally, the horizontal permeability of cohesive soils is several times higher than the vertical permeability, but the horizontal permeability was also shown to be reduced to the same value as the vertical permeability owing to the occurrence of the smeared zone. The smeared zone was also reported to be approximately 2–3 times the diameter of the casing at the time of drain casting and the permeability in this area decreases to approximately 1/5. Consequently, a permeability k' modified by the primary consolidation ratio may be used for convenience to account for the reduction in permeability due to the smeared zone. Hansbo [4] proposed a theoretical method to divide the analysis into two domains: a smeared zone with reduced permeability and an undisturbed zone without reduced permeability.

3.2.1.3 Execution

In the SD method, sand mats are placed on the surface of the cohesive soil layer as a drainage layer prior to the placement of drains. In several cases, the thickness of the layer is 1.0–1.5 m for marine works and 0.5–1.0 m for land works. In land works, the sand layer is not only used as a drainage layer, but also to ensure the trafficability of the execution machine, and the sand layer should be thickened as necessary. The general drain casting method is called the exclusion method, in which a casing pipe with a closed end is driven into the soft clay layer and sand is placed inside; then, the end of the casing pipe is opened and pulled out to the surface, leaving the sand to form a drain. The sand to be used as the drain material should be tested for suitability using a grain size test at each sampling site prior to the start of execution. If necessary, a permeability test should be conducted to check the permeability. To adapt to the severe environment of large water depths and weather/sea conditions, marine execution equipment incorporates a device to prevent the casing pipe from being shaken by the leader and to discharge water from the pipe by air pressure to ensure that sand is fed into the casing pipe. Currently, the Global Navigation Satellite System is the main method for managing casting positions, and automatic tide correction using tide gauges is also used for height management.

For PVD installation on reclaimed land, a sand mat of approximately 0.5–1.2 m is placed to ensure drainage on the ground and the trafficability of the execution machinery. A casing (mandrel) is generally used to place the drain. Recently, lightweight special-purpose machines based on hydraulic excavators have become mainstream. The friction roller method is the primary driving method, which enables low-vibration, low-noise static press-in process and can be used at great depths. For sites with limited height owing to the presence of high-voltage lines or air restrictions, a jointing method has also been developed, in which the casing is jointed and placed. Execution control items during the placement of drains include the placement depth, the volume of work, and photographs. It is also important to ascertain the amount of co-rise in execution management. The most important aspect to consider when placing a drain is to prevent the drain material from rising up with the casing when the casing is extracted, which is called 'co-lifting'. Factors that may cause co-lifting include mud flowing into the casing tip, anchor failure owing to an intervening stiff layer in the intermediate layer, and low shear resistance acting on the anchor owing to low ground strength. Because this can be addressed by selecting the appropriate tip anchor based on ground conditions, an appropriate anchor must be selected by performing a test execution.

When PVD installation is performed at sea, there are two methods: rigging the barge using a land machine or using a PVD installation vessel. The PVD installation method when the barge is outfitted with land machinery is almost the same as that for land installation; the difference is that the barge is moved and positioned with a winch and the cutting of the drain material is performed underwater. The PVD installation vessel is equipped with a float-type PBD method, which enables the PVD to be placed efficiently while a dedicated placement device traverses the rail. In combination with horizontal drains, the ground can be improved without the need for sand mats. When selecting the type of vessel, the barge type is often used in open seas where towing material to the execution location is possible, whereas the connected float type is commonly used in closed water areas where towing is not possible. Notably, in marine execution, the length of the casting varies depending on the tide level; therefore, a tide staff should be installed to control the level.

3.2.2 Vacuum Consolidation Using Vertical Drains

3.2.2.1 Overview

A vacuum consolidation method is used to reduce the water pressure in the drain by using the vertical drain method. When the water pressure in the drain is reduced, the pore water pressure in the cohesive soil becomes relatively high, and the pore water flows towards the drain material. Consequently, the pore water pressure in the cohesive soil decreases and the effective stress increases, resulting in consolidation. Because consolidation occurs isotropically as the pore water pressure decreases, theoretically, no shear forces are generated, which has the advantage that, unlike when overburden loads are applied by embankments, no slip failure or lateral displacement occurs. To obtain further consolidation effects, overburden loads, such as embankments, may be applied in combination, in which case, large consolidation settlements of 15–20 cm/day or more can be generated. Theoretically, the water pressure in the drain can be reduced to $101 \ kN/m_2$ (atmospheric pressure); however, in practice, the effective pressure reduction in the ground is 60–80 kN/m_2. This is due to the fact that cavitation, which creates voids in the water, occurs when the pressure is reduced beyond atmospheric pressure. Although this concept was introduced a long time ago, the technology was established only in the 1980s and has now become a method that is used worldwide.

There are two types of vacuum consolidation methods: sheets and caps. In the sheet-type vacuum consolidation method, the surface of the sand mat in the area to be improved is covered with an elastic airtight sheet to reduce the water pressure in the drain. The water pressure in the drain is also reduced by depressurising the airtight sheet using a vacuum pump. If a vacuum pump is used for both exhaust and drainage, the air under the airtight sheet is discharged, followed by the discharge of the groundwater that has flowed out. Another method is to install an air-water separation tank, where the water in the tank is discharged to the outside via a submersible pump and the pressure is reduced using a vacuum pump to drain the water. According to this method, stable depressurisation can be performed even if there is a large difference in height between the pump and the ground surface of the improved ground. Conversely, in the cap-type vacuum consolidation method, a cap is attached to the top of the drain and each drain is connected to a vacuum pump for direct drainage. It consists of a header pipe, water collection pipe, drainage hose, and capped PBD. Because this method does not require airtight sheets, it can also be used underwater, making it a useful method for ground improvement at sea. However, if the drain is in direct contact with water, the water is sucked up through it, and if it is in contact with air, it is sucked up through it. Consequently, this method can only be used when there is at least 1 m of ground with low permeability above the cap, and the cap is below the groundwater table. Furthermore, if there are highly permeable sand or gravel layers below the clayey layer, the drain must not be allowed to bottom out onto them. This is because the groundwater in the sand and gravel layers will be sucked up, and negative pressure will not act on the clayey soil. A separation of 2–3 m is advisable to allow greater leeway.

As the investigation and design of the vacuum consolidation method are generally similar to those of the PVD method, the details are as given in the previous section. Moreover, while considering the investigations, in the vacuum consolidation method, special attention must be focused on the pore water pressure conditions in the intermediate sand layer in the clayey layer and the ground below the clayey layer. If the presence of intermediate sand layers is predicted by the results of the borings, in situ permeability tests should be conducted as necessary. In terms of design, the depth of drain placement is, as described above, and the placement intervals are identical to that of the PVD method. Barron's solution can be used to determine the consolidation rate. For consolidation of the lower part of the improvement area, the amount and time of consolidation were considered, assuming one-dimensional consolidation in the vertical direction.

3.2.2.2 Execution

Figure 3.5 shows the procedure for the cap-type method as an example of the vacuum consolidation method. First, a preliminary survey was conducted before execution. The improvement area is divided into small blocks of approximately 3000 m² and soundings are conducted at the periphery of the improvement area and at the centre of the blocks. Here, 3000 m² is the area covered by one vacuum pump unit. In recent years, the use of multiple small, inexpensive vacuum pumps have also been adopted from an economic perspective. Then, the thickness of the clayey layer is estimated from the sounding results and drains are manufactured at the factory according to the length of the layer. The drains are then rolled and delivered to the site. The drains delivered to the site are driven into the ground using a PVD casting machine. If the work is to be conducted at sea, a barge or dedicated PVD installation vessel is prepared, the vessel is guided to the placement position, and the drain is placed at a predetermined location. A drainage hose is connected to the drain via the cap; thus, the drainage hose is connected to the water collection pipe, header pipe, and vacuum pump unit, and it is operated to reduce the water pressure in the drain.

The parameters to be controlled during execution include the applied negative pressure, amount of drainage, settlement, horizontal displacement at the periphery, and pore pressure in the ground. In addition to the value in the vacuum pump unit, the negative pressure is controlled either under an airtight sheet or at the top of the drain. Measuring the negative pressure in the vacuum pump unit confirms that the vacuum pump is working properly, whereas measuring the negative pressure at the airtight sheet or top of the drain confirms that the negative pressure is higher than the design value.

As an execution consideration in the cap-type vacuum consolidation method, the cap attached to the drain should be securely embedded in the cohesive soil to prevent leakage. The vacuum consolidation method draws in the ground around the area that is to be improved. If this displacement is a problem, steel sheet piles or deep mixing walls may be used

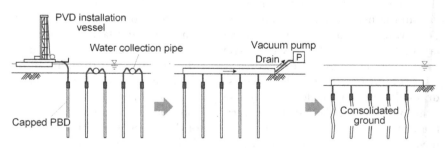

Figure 3.5 Execution procedure of the capped vacuum consolidation method.

in combination with the vacuum consolidation method. Moreover, naturally deposited ground often contains highly permeable layers, such as intermediate sand layers within the ground, and the negative pressure can reach the outside of the improvement area through the intermediate sand layers, causing settlement of the surrounding ground. The results of the preliminary investigations are used to determine the sediment distribution in the intermediate sand layer, and a water barrier seal is attached to the intermediate sand layer to prevent the transmission of negative pressure when the drains are manufactured at the factory. Impermeable seals are usually installed up to 1 m above and below the intermediate sand layer.

3.3 COMPACTION METHODS

3.3.1 Sand Compaction

3.3.1.1 Overview

The vertical drain method is used to promote consolidation to increase the stiffness and strength of soft cohesive soils. The compaction method is a ground improvement method that promotes consolidation of soft clayey soils to a greater extent than the vertical drain method or compaction of soft sandy soils to improve stiffness and strength. The vertical drain method is used to promote consolidation and targets soft clayey soils, whereas the compaction method targets a wide range of soils, from soft clay to sandy soils. The purpose of compaction for clay is to increase the stiffness and strength; however, compaction for sandy soils also includes a liquefaction countermeasure. Various compaction methods are available, but the sand compaction pile (SCP) method is introduced first. In this method, large-diameter, well-compacted sand piles are placed in the ground. The casing pipe (steel hollow pipe) is driven into the ground via the vibrations from a vibro-hammer and rotational force from a rotating device, and the sand is left behind when the casing pipe is pulled out and driven back in. This process is repeated to form a compacted sand pile in the ground. This is a highly reliable method that can consistently drive sand piles larger than the diameter of the casing pipe. When high-quality sand is difficult or uneconomical to obtain or when high strength is expected from the sand pile itself, gravel may be used instead of sand as an improvement material. For more information on the SCP method, please refer to a previously published book [5], which provides an elaborate summary of the details of the SCP method.

The SCP method was developed in Japan and has been used at both sea and land since the 1960s, making it one of the most widely used ground improvement methods in Japan. It is also used in East and South-East Asia. The principle for improving the SCP method for clay soils is to form a composite ground by constructing compacted sand piles in soft clay soils. The stiffness and strength of the sand pile, which is greater than that of the cohesive soil, combined with the consolidation effect of the surrounding cohesive soil because of the permeability of the sand pile, can increase the bearing capacity for the overburden load and shear resistance. It can also reduce the settlement and lateral displacement. It is possible to forcibly replace soft cohesive soils with sandy soils by means of a high replacement ratio of more than 70%. The SCP method for sandy ground is primarily based on the principle that vibrations caused by compaction promote the rearrangement of sand particles and simultaneously, the circular spreading of the pile shears the ground and makes it denser. Pressing sand piles into the ground also increases the lateral confining pressure. These effects render soft sandy soils that are less prone to liquefaction during earthquakes.

The primary improvement applications of the SCP method are shown in Figure 3.6. The range of the replacement rate a_s and purpose of the improvement are appended to the figure. a_s is the ratio of the volumes of the sand pile and original ground. Sand piles can be placed in various configurations, such as square, triangular, and rectangular formations. Another feature of the SCP method is that it can be applied at great depths: approximately 45 m in land execution and 70 m below the surface under water using marine execution vessels. The standard diameter of sand piles after driving is $d = 0.7$ m for land works, while $d = 1–2$ m is often used for marine works. As vibro-hammering is used in the normal SCP method, there are concerns about the effects in the surrounding areas, such as vibration, noise, and displacement of adjacent structures; therefore, it is important to investigate the surrounding conditions and implement measures to prevent problems before application. If vibration or noise is a problem, a rotary press-in method for casing pipes can be used instead of using vibro-hammers.

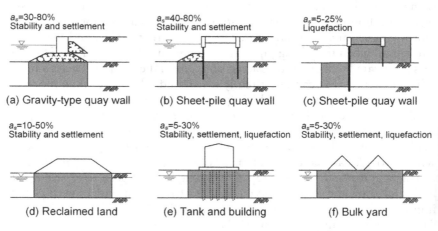

a_s=30-80%
Stability and settlement

a_s=40-80%
Stability and settlement

a_s=5-25%
Liquefaction

(a) Gravity-type quay wall   (b) Sheet-pile quay wall   (c) Sheet-pile quay wall

a_s=10-50%
Stability and settlement

a_s=5-30%
Stability, settlement, liquefaction

a_s=5-30%
Stability, settlement, liquefaction

(d) Reclaimed land   (e) Tank and building   (f) Bulk yard

Figure 3.6 Applications of SCP method.

3.3.1.2 Investigation

The required geotechnical investigations vary significantly depending on the purpose for ground improvement (e.g., stability improvement, settlement control, bearing capacity improvement, and liquefaction control) and soil type (e.g., clayey or sandy soil). In the case of clayey soils, in addition to soundings and borings, physical, consolidation, and shear tests must be conducted on the sampled soil. The results of consolidation tests are used to estimate the final settlement and consolidation time, whereas shear tests are used to estimate the increase in strength. In the case of sandy soils, geophysical tests to determine the grain size and SPTs to determine N values are necessary. This is because the liquefaction strength is usually evaluated based on N values, and the improvement effect is also evaluated based on the increase in N values. When the effectiveness of ground improvement is evaluated using finite element analysis, more detailed investigations, such as undrained cyclic triaxial tests to investigate liquefaction characteristics, are required.

The physical properties of the improvement material used in the SCP method should also be investigated. The grain size additive curves of the proven materials are shown in Figure 3.7, where the fine fraction with grain sizes smaller than 0.075 mm is less than 5%, and the maximum grain size should be less than 40–50 mm. Moreover, the soil should not become finer during compaction. If such materials are not cheaply available at the site, strength and workability tests may be performed on available materials, and the results may be incorporated into the design and execution. In high

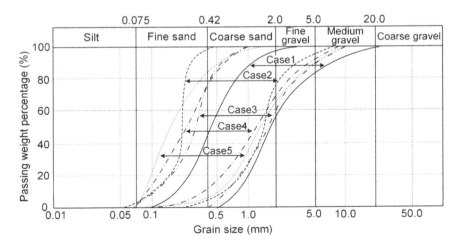

Figure 3.7 Grain size distribution curve of soil used in the past. (Adapted from Ref. [1].)

replacement improvements, where no drainage effect is expected, materials with 10–15% fine fraction have been used in some cases. Furthermore, in recent years, recycled materials, such as steel slag and oyster shells, have been actively used as improvement materials.

3.3.1.3 Design for Clayey Ground

The design methods differ significantly depending on whether the soil to be improved is clayey or sandy. This is not surprising, because the purpose of the improvements is different in each case. The design of the SCP method for cohesive soils is described first. When clayey soils are to be improved, the objective is to increase the stiffness and strength of the soil, increase the stability of coastal structures, and reduce displacement. The first step is to investigate the increase in shear strength due to this improvement. The ground composed of the original cohesive soil and sand piles is referred to as a composite ground, and a conceptual diagram of this is shown in Figure 3.8. When the composite ground is subjected to overburden loads, for example, from reclaimed soil, the stresses are concentrated on the sand pile because of the difference in stiffness between the clay and the sand pile. The ratio of the stresses acting on the sand pile σ_s to the stresses acting on the clayey soil σ_c is called the stress sharing ratio n and is an important parameter in the design. If the replacement ratio is a_s, σ_s and σ_c are calculated using the following equations.

$$\sigma_s = n / \left\{ 1 + (n-1)a_s \right\} \cdot \sigma \tag{3.5a}$$

$$\sigma_c = 1 / \left\{ 1 + (n-1)a_s \right\} \cdot \sigma \tag{3.5b}$$

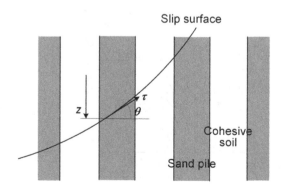

Figure 3.8 Composite ground composed of clay and sand piles.

There are various arguments regarding the stress sharing ratio n, but according to inverse analysis from field data, it is considered to be in the range of 2–6. However, in the design, n is set as follows to ensure safety of the design.

$$a_s < 0.4 \qquad : \quad n = 3, \quad \phi_s = 30° \tag{3.6a}$$

$$0.4 \le a_s < 0.7 \qquad : \quad n = 2, \quad \phi_s = 30 - 35° \tag{3.6b}$$

$$0.7 \le a_s \qquad : \quad n = 1, \quad \phi_s = 35° \tag{3.6c}$$

The stability of the entire coastal structure, such as a quay wall or seawall, and the bearing capacity of the improved ground to the applied loads can be assessed by circular slip analysis using the modified Fellenius method. The shear strength τ of the composite ground can be evaluated using the following equation:

$$\tau = (1 - a_s)\left\{c_0 + kz + \Delta\sigma_z\mu_c\left(\Delta c / \Delta p\right)U\right\} + \left(w_s z + \mu_s\Delta\sigma_z\right)a_s \tan\phi_s \cos^2\theta, \tag{3.7}$$

where c_0 is the undrained shear strength at the ground surface, k is the rate of increase in strength with depth, z is the depth, $\Delta\sigma_z$ is the mean value of the vertical stress increment, μ_c is the stress reduction factor for the cohesive soil $\left(1 / \left\{1 + (n-1)a_s\right\}\right)$, $\Delta c / \Delta p$ is the strength increase rate of the cohesive soil, U is the mean degree of consolidation, w_s is the unit weight of the sand pile, μ_s is the stress concentration factor for the sand pile $\left(n / \left\{1 + (n-1)a_s\right\}\right)$, ϕ_s is the internal friction angle of the sand pile, θ is the angle between the shear plane and the horizontal plane. As shown in this equation, the shear strength of the original soil, strength increase ratio, improvement ratio, shear strength of the sand pile, and its own weight are considered. σ_z in the equation is the vertical stress increment due to the overburden load; however, stress dispersion should be considered, for example, by using the Boussinesq equation. If the stability of the structure as a whole and the bearing capacity of the applied load cannot be ensured, then the shear strength should be increased by increasing the replacement ratio. Moreover, the SCP method is used to reduce the principal earth pressure on retaining structures and improve the passive earth pressure on sheet piles. As the applicability of the SCP method to these problems has not yet been fully investigated, it is desirable to assess its stability using various proposed equations and numerical analyses.

When structures are built after SCP improvements, immediate deformations associated with shear and long-term consolidation settlement occur. For two- or three-dimensional shear deformations, estimating the displacements with simple models is difficult and numerical analysis using finite element analysis is required. Various modelling approaches have

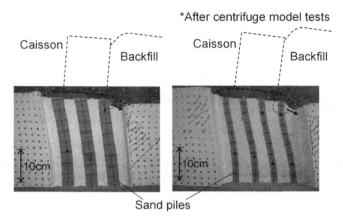

Figure 3.9 Failure pattern of composite ground composed of clay and sand piles. (Based on Ref. [6].)

been proposed for modelling SCP improvement sections in numerical analysis. In the case of high replacement rates, the improved ground is almost entirely sandy; it is evaluated as a uniform sandy soil, and a reduction in strength corresponding to the replacement rate is expected. In the case of low replacement rates, it is known that the composite ground of sand piles and cohesive soils does not slip and fail in the same manner as a uniform ground, as shown in Figure 3.9; therefore, a model with alternating layers of sand and cohesive soils should be used [6].

For the consolidation settlement, the final consolidation settlement is determined by multiplying the settlement as no treatment determined on the basis of soil tests by the settlement reduction factor β. β is a function of stress concentration effects and the rate of replacement and is expressed as follows:

For a low replacement ratio $\left(a_s < 0.5\right): \beta = 1 / \left\{1 + (n-1)a_s\right\},$ (3.8a)

For a high replacement ratio $\left(a_s > 0.5\right): \beta = 1 - a_s.$ (3.8b)

Settlement control during execution is important because of the large variation in settlement. Similar to the drain method, the consolidation settlement rate is calculated from Barron's solution based on the sand pile diameter, spacing, and arrangement (square or triangular arrangement). It is known that as the replacement ratio increases, the consolidation velocity becomes 2–10 times slower than that of Barron's solution; however, the layer thickness of cohesive soil at high replacement ratios is thin, and the reduction in consolidation velocity is not a significant practical problem.

In marine execution, where the ground is soft and generally improves with a high replacement rate, the underlying ground is raised as the SCP is placed. In large cases, the raising height can be several metres. Several formulae have been proposed to estimate the rise in height, for example, the following equations are used in the design:

Diameter of sand pile is less than 2.0 m : $\mu = 0.356a_s + 2.803L^{-1} + 0.112,$

$$(3.9a)$$

Diameter of sand pile is 2.0 m : $\mu = 0.718a_s + 2.117L^{-1} + 0.056,$ (3.9b)

where μ is the upheaval ratio (ratio of the amount of upheaval to the sand piles that are cast) and L is the length of the sand pile. The actual shape of the upheaval depends on the driving sequence, and studies based on field data are in progress. It is also possible to continue the improvement and drive SD or SCP without removing the fill. In cohesive soils, the strength of the inter-pile cohesive soil is reduced by the disturbance caused by SCP placement; however, the strength reduction is often not considered in the normal design. This is because the strength loss associated with a disturbance recovers within 1–3 months. In cases where structures are to be installed immediately after sand pile driving, a reduction in strength should be considered in the design.

3.3.1.4 Design for Sandy Ground

To address the liquefaction problem, which is one of the challenges of the soft ground, the SCP method is applied to sandy soil. Figure 3.10 shows the basic concept of improving sandy soil using the SCP method, which

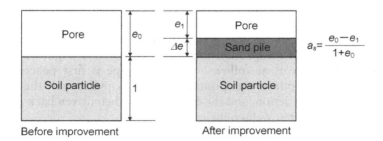

Figure 3.10 Configuration of sandy ground improved by sand compaction pile (SCP) method.

involves vibratory injection of sand equivalent to Δe into a ground with a volume of $1+e_0$ (where e_0 is the original ground void ratio). As described in Chapter 2, the liquefaction strength of the ground and improvement effect are evaluated by the N value obtained from the SPT. The N value after improvement is calculated in relation to the N value of the original ground before improvement (N_0) and the replacement ratio a_s; however, it is also influenced by other factors such as the grain size distribution of the original ground and soil overburden pressure. In particular, if the ground contains a large amount of fine fraction of 0.075 mm or less, the improvement effect will be smaller; therefore, correction is necessary.

There are four design methods: using a simplified diagram based on actual results (Method A) and estimating the pore ratio from the N value via the relative density D_r (Methods B, C, and D). Design methods have been improved from Methods A to D, and the design is basically based on Method D. Methods A to C are regarded as simplified methods and are discussed in another book [5]. Method D is introduced in this section. The design flow and model of the ground after driving the sand piles are shown in Figure 3.11, where the reduction in the improvement effect can be considered, such as the reduction due to fine fractions and the reduction due to the raising of the ground caused by the driving of sand piles. As shown in the figure, a parameter defined as the effective compaction factor R_c is introduced to consider changes in the ground after sand pile driving. By relating R_c to the fine fraction content F_c of the ground, the post-improvement pore ratio e_1 and relative density D_{r1} can be reasonably assessed. e_1 is obtained by calculating $e_0 - R_c (1+e_0) a_s$. The relationship between F_c and R_c has also been investigated [5] and is a design method that can be used for soils with a relatively large fine fraction content. Another design method using the cumulative shear strain of the soil has also been proposed [5].

3.3.1.5 Execution

Examples of marine execution vessels and fleet configurations for the SCP method are shown in Figure 3.12. The approximate dimensions of the execution vessels are also shown. In reality, there are vessels of various sizes. There are several types of SCP methods; however, this section describes a typical 're-drive and compaction method'. As shown in Figure 3.13, the execution procedure is as follows: the casing pipe is first penetrated to a predetermined depth in the ground and then is pulled out; the sand is discharged with this action, and the casing pipe is then driven back again to create a large-diameter, well-compacted sand pile.

As control items during execution, care must be taken to ensure the quality of the improvement sand used, depth of the seabed before and after improvement, sand pile length, amount of improvement sand applied, sand pile driving position, and strength of the sand pile. For example, depth

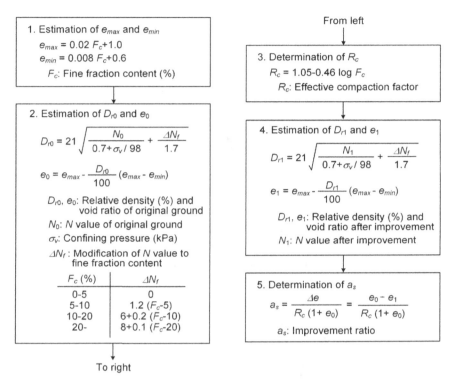

1. Estimation of e_{max} and e_{min}

$e_{max} = 0.02 \, F_c + 1.0$

$e_{min} = 0.008 \, F_c + 0.6$

F_c: Fine fraction content (%)

2. Estimation of D_{r0} and e_0

$$D_{r0} = 21 \sqrt{\frac{N_0}{0.7 + \sigma_v / 98} + \frac{\Delta N_f}{1.7}}$$

$$e_0 = e_{max} - \frac{D_{r0}}{100} (e_{max} - e_{min})$$

D_{r0}, e_0: Relative density (%) and void ratio of original ground

N_0: N value of original ground

σ_v: Confining pressure (kPa)

ΔN_f : Modification of N value to fine fraction content

F_c (%)	ΔN_f
0-5	0
5-10	1.2 (F_c-5)
10-20	6+0.2 (F_c-10)
20-	8+0.1 (F_c-20)

From left

3. Determination of R_c

$R_c = 1.05 - 0.46 \log F_c$

R_c: Effective compaction factor

4. Estimation of D_{r1} and e_1

$$D_{r1} = 21 \sqrt{\frac{N_1}{0.7 + \sigma_v / 98} + \frac{\Delta N_f}{1.7}}$$

$$e_1 = e_{max} - \frac{D_{r1}}{100} (e_{max} - e_{min})$$

D_{r1}, e_1: Relative density (%) and void ratio after improvement

N_1: N value after improvement

5. Determination of a_s

$$a_s = \frac{\Delta e}{R_c (1 + e_0)} = \frac{e_0 - e_1}{R_c (1 + e_0)}$$

a_s: Improvement ratio

To right

Figure 3.11 Design procedure for sandy ground using Method D.

surveys in marine works should be performed within a range of 1.5–2.0 times the pile length from the edge of the improvement area to cover the area affected by the embankment. In SPTs for sandy soil, the fine fraction, which affects the improvement effect, should be measured in the collected samples. Sand quantity control is also important. Sand piles are formed by confirming that sand equivalent to the length of the sand pile to be formed has been removed when the casing pipe is withdrawn using a sand level gauge and is then driven back to the prescribed height. The ratio of the amount of sand to be delivered to the site to the designed sand input is referred to as the required additional ratio. This includes the volume change due to vibratory compaction and execution losses, which have been proven to be generally around 1.3 and 1.0–1.05, respectively. The required additional ratio is approximately 1.4 on land and 1.4–1.45 at sea. Another important aspect in the execution of SCP methods is that it is necessary to build sand piles to a specified depth. The length of the sand pile can be ascertained using a depth gauge; however, when the sand pile is securely

(a) Whole view

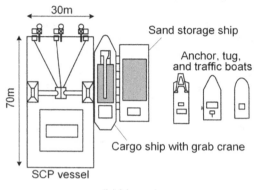

(b) Vessels

Figure 3.12 SCP vessel and fleet composition. (Courtesy of Fudo Tetra Co.)

placed in the bearing stratum, it is checked by the penetration speed of the casing pipe. The objective value of the control is obtained by performing test execution at the ground investigation site and ascertaining the relationship between the penetration speed and the bearing stratum.

Another aspect to be noted is that relatively large vibrations and noise are generated during execution, particularly in methods using vibro-hammers. Moreover, as materials such as sand are forcibly injected into the ground, there is a high possibility of deformation of the surrounding ground or structures during execution. Because there are prior measurement results for the vibration, noise, and displacement of the surrounding ground, the execution must be planned with reference to these results.

For quality control, it is common practice to check the N value obtained from the SPT after improvement to confirm the degree of compaction. If the work is offshore, a depth survey is implemented to check the rise of the ground surface.

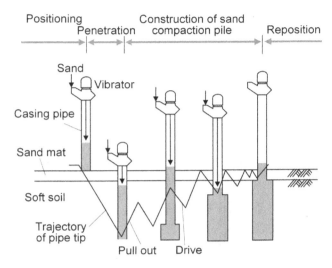

Figure 3.13 Execution procedure of SCP.

3.3.2 Vibro-compaction

3.3.2.1 Overview

The vibro-compaction method (including the vibro-flotation method) and vibro-replacement method (also known as the vibro-stone column) are the most commonly used ground improvement methods worldwide. The vibro-compaction method is a vibratory compaction method for loose sandy soil, which uses horizontal vibration and the water tightening effect to increase the stiffness and strength of the ground. It is sometimes used as a liquefaction countermeasure to compact sandy soil through vibration. Sand and gravel are used as improvement materials. This ground improvement method was developed in Germany in the 1930s and subsequently used in the USA, where it has a history of nearly 100 years. Conversely, the vibro-replacement method is a vibratory compaction method that is primarily used for cohesive soils with undrained shear strengths of 15 kN/m$_2$ and above, where a gravel pile is formed in the ground with crushed stone or concrete shells; for cohesive soils with a shear strength less than 15 kN/m$_2$, a gravel pile encased with geosynthetic-encased granular columns is used. This method involves driving gravel piles in a compact manner, which increases the average strength of the combined gravel pile and cohesive soil owing to the higher shear strength of the gravel pile. Although this method may not be regarded as a compaction method for cohesive soils because it does not actively increase the pile diameter as in the SCP method, it is

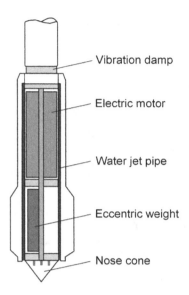

Vibration damp

Electric motor

Water jet pipe

Eccentric weight

Nose cone

Figure 3.14 Vibrators used for vibro-compaction.

classified as a compaction method in this book from the perspective of the compaction of gravel piles. This method was also developed in European countries and is now one of the ground improvement techniques used worldwide.

In these methods, a rod-shaped vibrating body with a built-in eccentric load, called a vibrator, as shown in Figure 3.14, is used; the vibrator, which is suspended from a crane, is penetrated to a predetermined depth by its own weight and horizontal vibration is caused by the rotation of the eccentric load. Water (or a mixture of water and compressed air) is injected from the tip of the vibrator in the vibro-compaction method. While extracting the vibrator, improvement material (sand or gravel) is injected from the ground surface or tip of the vibrator to form sand or gravel piles, and the vibrations compact the improved piles and surrounding ground. Vibratory compaction using a high-frequency vibrator, which requires relatively little energy, enables low-vibration and low-noise operations. These methods are expected to increase the vertical bearing capacity and reduce settlement as a composite ground by forming sand and gravel piles that are filled with the improvement material. The vibro-compaction method is primarily used for improvements at depths that are shallower than 20 m, while the vibro-replacement method is predominantly used for improvements at depths shallower than 15 m. Both marine and land machines are available. For more information on the vibro-compaction and vibro-replacement methods, please refer to previously published books [7, 8].

The vibro-compaction and SCP methods for sandy soils commonly use sand and gravel to compact the ground. The most significant difference is that the vibro-compaction method supplies sand or gravel to the voids created in the ground by vibration, whereas the SCP method supplies sand and compaction after the casing pipe is pressed into the ground while vibrations are applied. Therefore, in the vibro-compaction method, the amount of improved material and the extent of influence of the compaction vary depending on the ground conditions, vibrator specifications, capacity, and execution method. It is also argued that, when gravel piles are formed in sandy ground, liquefaction countermeasures may be expected; however, the drainage effect cannot be expected when the soil enters the gravel pore spaces during execution. The effectiveness of liquefaction countermeasures owing to drainage awaits the results of post-earthquake investigations and other actual results. Moreover, the SCP method uses a casing pipe for execution, which enables improvements to be achieved at large depths.

3.3.2.2 Investigation and Design for Sandy Ground

The vibro-compaction method is used to compact soft sandy soils, but it is important to understand the physical properties of the original ground, including grain size, as part of the ground investigation required for its design. In particular, determining the content of fine fraction is important. This is because it is difficult to apply this method to soils with a high fine fraction content, and sandy soils with a fine fraction of less than 10% (0.075 mm or less) and a maximum of 20% or less should be considered for improvement. Moreover, the clay content should not exceed 2–3%. The upper limit for fine fraction is set at 20% because it was found that compaction is not effective if the fine fraction content is higher than this. It is also necessary to conduct a SPT or cone penetration test (CPT) on the original ground to determine the current degree of compaction and to set the target degree of compaction. These investigations are necessary to determine the stratigraphic composition.

To compact soft sandy soils and increase the stiffness and strength of the ground, the degree of densification must be considered in the design. The procedure involves first determining the void ratio and relative density of the original ground. These are usually estimated from the SPT and CPT results. Subsequently, the shear strength and liquefaction strength of the ground required by the structure to be built on the improved ground are established, and the target void ratio and relative density are determined to achieve these strengths. For example, a relative density of 60% or higher can be set in each guideline for embankment structures. Finally, the settlement and soil supply are calculated to achieve the void ratio and relative density, and execution is performed to satisfy these requirements. If no soil is supplied, the theoretical settlement S is obtained from the following equation:

$$\frac{S}{h} = \frac{e_0 - e_1}{1 + e_0},$$

(3.10)

where h denotes the depth of improvement, e_0 denotes the void ratio of the original ground, e_1 denotes the target void ratio of the improved ground. Thus, assuming that the height of the ground surface does not change with the supply of the improvement material, the amount of improvement material V required in the square or equilateral triangle bounded by the pile core is as follows:

Square arrangement: $V = s^2 h \dfrac{e_0 - e_1}{1 + e_0},$

(3.11a)

Equilateral triangle arrangement: $V = \dfrac{\sqrt{3}}{4} s^2 h \dfrac{e_0 - e_1}{1 + e_0},$

(3.11b)

where s is the distance between the piles. If the diameter of an improved pile is d_w, the volume of one pile is $\pi d_w^2 h / 4$; therefore, by substituting V in Eq. (3.11), s can be obtained as follows:

Square arrangement: $s = 0.89 d_w \sqrt{\dfrac{1 + e_0}{e_0 - e_1}},$

(3.12a)

Equilateral triangle arrangement: $s = 0.95 d_w \sqrt{\dfrac{1 + e_0}{e_0 - e_1}}.$

(3.12b)

These equations are theoretical and often do not match the results of the actual execution. Various charts based on actual experience have been proposed, which should be used in conjunction with these charts; furthermore, it is important to compensate for this by means of test execution.

3.3.2.3 Investigation and Design for Clayey Ground

In addition to soundings and borings, geophysical, consolidation, and shear tests on the sampled soil are necessary for the design of the vibro-replacement method. The results of the consolidation tests are used to estimate the final settlement and consolidation time, while shear tests are used to estimate the increase in strength. The physical properties of the gravel used for gravel piles should also be investigated. Grain sizes of 50%, 20%, and 10% (D_{50},

D_{20}, D_{10}, respectively) should be used to obtain the suitability number S_N using the following formula.

$$S_N = 1.7 \sqrt{\frac{3}{\left(D_{50}\right)^2} + \frac{1}{\left(D_{20}\right)^2} + \frac{1}{\left(D_{10}\right)^2}} \qquad (3.13)$$

$S_N < 10$ denotes superior suitability, whereas $S_N > 50$ is unsuitable; a lower value indicates greater suitability of the material for gravel piles.

The objective is to increase the stiffness and strength of the cohesive soil to increase the stability and reduce the displacement of coastal structures. The design will involve estimating the strength and deformation of the composite ground composed of the original cohesive soil and gravel piles. The general arrangement and pile diameters of the gravel piles are described subsequently. In addition to square and triangular arrangements, there are also radial arrangements with larger distances between the piles. This arrangement is used for tank foundations, for example, where the strength of the central part of the pile must be increased. Pile diameters of 0.5–1.2 m are common. Next, the design method for the bearing capacity is explained. Based on theoretical considerations, the bearing capacity of the gravel pile $q_{ult,g}$ is expressed by the following equation:

$$q_{ult,\ g} = K'K_p c_u, \qquad (3.14)$$

where K' is the coefficient for the anticipated effect of lateral stresses on the gravel pile due to overburden loads, K_p is the passive earth pressure coefficient, c_u is the undrained shear strength of the original ground. Different values have been suggested by various researchers for $K'K_p$ and it is assigned values between 12 and 25. An approximate estimate of 20 was used. If the bearing capacity of the original ground is $q_{ult,c} = 5c_u$, then the bearing capacity of the composite ground q_{ult} is given by the following equation:

$$q_{ult} = q_{ult,g} a_s + q_{ult,c} \left(1 - a_s\right), \qquad (3.15)$$

where a_s is the replacement rate by gravel piles.

When considering the stability of the entire structure, including the improved ground, the shear strength of the composite ground is considered; however, considering the composite ground as a uniform ground, the undrained shear strength c_{eq} and the internal friction angle ϕ_{eq} of the ground are averaged with the replacement ratio as follows:

$$c_{eq} = c_c \left(1 - a_s\right) \qquad (3.16a)$$

$$\phi_{eq} = \tan^{-1}\left\{a_s \tan\phi_g + \left(1 - a_s\right)\tan\phi_c\right\}, \tag{3.16b}$$

where c_c is the undrained shear strength of the underlying soil, ϕ_g and ϕ_c are the internal friction angles of the gravel pile and original ground, respectively. As this equation does not consider the effect of stress concentration on the gravel pile, the following correction parameter ω_f may be included in the a_s of Eq. (3.16b):

$$\omega_f = \frac{a_s n}{1 + a_s\left(n - 1\right)}, \tag{3.17}$$

where n is the stress sharing ratio and ω_f is considered to be in the range 0.4–0.6.

There are several methods for estimating the final consolidation settlement, including the stress reduction, improvement factor, and elastic-plastic methods. The stress reduction method, which has been used since the earliest times, is explained by the following equation for the consolidation settlement S' of a composite of the original ground and gravel piles:

$$S' = \frac{1}{\beta}S = \frac{1}{1 + a_s\left(n - 1\right)}S = \frac{1}{1 + a_s\left(n - 1\right)}m_v \Delta\sigma_z h, \tag{3.18}$$

where S is the consolidation settlement of the original ground without improvement, m_v is the consolidation coefficient of the original ground, $\Delta\sigma_z$ is the mean value of the vertical stress increment, h is the depth of the original ground. The settlement of the improved ground is calculated by dividing the consolidation settlement of the unimproved settlement by the coefficient β. Various equations have been proposed for the consolidation rate, derived from Barron's solution described for the vertical drainage method. The design concept of the vibro-replacement method considers the effect of the rigidity of the gravel pile in addition to the drainage effect. Owing to the presence of gravel piles in the ground with greater stiffness than the original ground, consolidation of the composite ground will theoretically be faster than consolidation based on drainage effects alone. Conversely, it is also known that the consolidation is slower than the theoretical value because the driving of the gravel piles disturbs the original ground and some of the gravel piles become clogged. The consolidation rate should be considered while considering these factors.

3.3.2.4 Execution

The vibrator and other equipment should be selected according to the ground improvement objectives, spacing between gravel piles, and depth

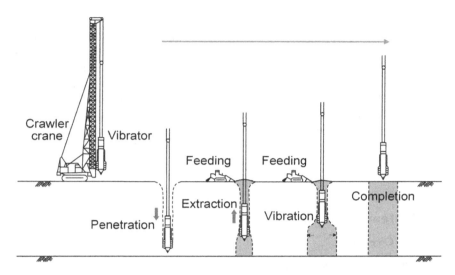

Figure 3.15 Execution procedure of the vibro-compaction method.

of improvement. The equipment required for this method includes a hydraulic or electric vibrator, crane with a lifting capacity of approximately 200 kN, tyre shovel, hopper, water pump, sand pump, water reservoir, and compressor. An overview of the vibro-compaction method is shown in Figure 3.15. The vibrator is moved to the improvement position and allowed to sink into the ground at a speed of approximately 0.3–0.5 m/min. The current value of the vibrator is measured, and the penetration speed is reduced if the current is too high. It is also important to reduce the water-jetting force when it reaches a layer which should not be disturbed. When the lower end of the improvement area is reached, the vibrator is allowed to vibrate for at least 30 s and then raised again at a speed of approximately 0.3–0.5 m/min. At every 0.3–0.5 m of raising height, the pulling should be temporarily stopped, and 30 s of vibration should be applied. In some cases, penetration is used to further compact the ground. If the ground settles significantly because of compaction, the vibrator is fed with a material (sand or gravel) either from the ground surface or from the end of the vibrator. This process is repeated until the ground surface is reached, thus completing the improvement in that area.

The execution of the vibro-replacement method is similar, but differs in that the vibrator is raised and lowered twice when it reaches the lower end of the improvement area; moreover, gravel or other improvement material is fed from the tip of the vibrator every 0.6–1.2 m of extraction to form an improvement pile. The vibrator is also used for the compaction of the improvement pile by penetrating the vibrator to a predetermined depth each time the improvement material is fed into the pile.

A point to note during the execution is that the vibro-replacement method discharges mud as a slurry; consequently, a pond should be prepared to collect it. The vibro-compaction and vibro-replacement methods are widely used and have a wide range of applications, but the improvement effect varies significantly depending on the soil layer composition and soil condition. It is necessary to revise the improvement specifications according to execution management, rather than relying solely on past results. For quality control, the N value by SPT or cone penetration value by CPT is checked after improvement to confirm that the prescribed degree of compaction has been achieved.

3.3.3 Heavy Tamping

3.3.3.1 Overview

A method of compacting soft sandy soils that has been used worldwide for several years is the heavy tamping method. In this method, a steel or concrete dead weight (generally weighing 50–300 kN) is lifted by a crane and dropped from a height (5–30 m) onto the ground surface; thus, impact forces are repeatedly applied to compact and strengthen the ground. The ground is compacted and strengthened by repeatedly applying impact forces. The ground can be compacted to a depth of up to 15 m. Large weights of 500 or 1700 kN are sometimes used as weights. The ground improvement method of compacting the ground by dropping weights has been used for a long time and was technically established by Louis Menard in France in the late 1960s. It can be applied to a wide range of soil types, such as rock fill, sandy soil, clay, waste, and peat, and is used not only for coastal structures but also for various facilities (e.g., roads, railways, electric power facilities, oil tanks, and plant foundations) to increase and equalise the bearing capacity, reduce residual settlement, and prevent liquefaction in the event of earthquakes. However, compacting the ground underwater via tamping is difficult; it is generally used for ground improvement works on land. Owing to the noise and vibration involved, it is difficult to use this method in areas if there are other existing structures, or if the site is close to an urban area. The results of various studies on noise and vibration are presented in the guidelines.

3.3.3.2 Investigation and Design

The ground investigation required for the design of the heavy tamping method is to determine the grain size by geophysical testing. This method is not suitable for soils with a high fine fraction content because the ground is compacted by dynamic energy. A SPT or CPT should be performed on the original ground to determine the current degree of compaction and set a target degree of compaction.

The process of repeatedly dropping weights into the ground is called tamping and can be divided into main tamping, which improves the deeper parts of the ground, and finishing tamping, which improves the shallower (surface) parts of the ground and is performed as a subsequent process. The main tamping design is performed as follows. First, the weight of the dead weight and height of the fall are considered. It is known that the weight of the dead weight W (kN) and height of fall H (m) are related to the depth of improvement D (m) and have the following relationship:

$$D = m\sqrt{WH / g},\tag{3.19}$$

where m is the reduction factor of the improvement effect and g is the acceleration due to gravity. Initially, D was said to be equal to $\sqrt{WH / g}$, but as experience was gained, the depth of improvement was determined to be shallower and a correction was required. This correction factor is denoted as m in the equation. In practice, D increases non-linearly with $\sqrt{WH / g}$. As the density of the ground of interest increases, D becomes smaller. Empirically, $m = 0.9$ for relatively shallow improvements in loose sandy soils and m is ~0.25 for deeper improvements. On average, $m = 0.5$ is often used. In any case, as the values of W and H increase, the improvement becomes more efficient. However, there are limits to the lifting capacity of the crane used. For example, in Japan, weights of 50–120 or 200–300 kN are selected and adjusted by the drop height if necessary. The method adjusts the drop height as required. Note that applying too much impact energy to the ground may cause the surrounding ground to rise and prevent consolidation.

The next consideration is the spacing between dropping points and number of phases. Instead of executing from the edges in sequence, the weights are dropped evenly across the plane in several phases (usually around three phases). In the first phase, the spacing should be equal to the depth of improvement. For example, if the depth of improvement is 8 m, then the dropping point spacing should be 8 m × 8 m. In rock fill or waste ground, the spacing may be marginally narrower, as the dropping effect is more pronounced directly downwards; in the second phase, the midpoints between the dropping points of the first phase are struck. Thus, the midpoints are struck in phases, and all points are struck evenly across the plane. The square of the smallest dropping point interval l_{min} at the end is the shared area per blow A. When divided into three phases, l_{min} is often half the dropping point interval of the first phase, that is, half the improvement depth.

Finally, the number of blows N_B at each point is set. The weight W multiplied by the height of fall H is the dropping energy, and the energy per unit area is $E = WH/A$. Multiplying this by the number of blows N_B at each dropping point presents the total impact energy per unit area E_t, which is expressed by the following formula:

$$E_t = EN_B = \frac{WH}{l_{min}^2} N_B. \tag{3.20}$$

Based on the empirical relationship between the total impact energy and the degree of compaction of the ground, the required impact energy per unit area is set and N_B is determined. For example, in Japan, the difference ΔN between the pre- and post-improvement N values is set and the required impact energy per unit volume of ground E_v for this amount of densification is determined from an empirical chart. In practice, $E_v = 300–600$ kN·m/m^3 has been designed in many cases, and in the case of waste volume reduction, a larger value may be considered as $E_v = 400–800$ kN·m/m^3. Multiplying this by the improvement depth D provides the total impact energy per unit area required, E_t, which is substituted into Eq. (3.20) to determine N_B.

When the ground with a high water content is to be improved, tamping generates excess pore water pressure, which takes time to dissipate. As the compaction effect cannot be achieved by continuing the tamping process when the pore water pressure has increased, it is necessary to wait for the excess pore water pressure to dissipate by allowing a settling period. A period of 1–2 days is required for sandy soils and 1–2 weeks for sandy silt soils.

Finishing tamping is designed to compact the shallow part of the ground (to a depth of 2–3 m) in and around the loosened borehole after the main tamping operation using less energy per blow than for the main tamping operation, with a weight of 50–120 kN and drop height of 5–10 m. Tamping is performed by tamping the entire ground surface.

3.3.3.3 Execution

In addition to tamping operations, it is important to combine execution management. Large crawler cranes are used for execution; however, to efficiently and repeatedly drop the dead weight from a predetermined height, cranes that are compatible with the dead weight-drop compaction method must be used, with improved hoisting power (winch), reinforced clutch and brake parts, and impact resistance to various structural parts of the body. In execution, owing to the lifting capacity of the crane, a single-suspension drop method is used for weights up to 120 kN and a detachment drop method using a hydraulic chucking device or similar is used for weights greater than 120 kN. Figure 3.16 shows schematic diagrams of standard tamping operations. Tamping creates blowholes in the ground with a diameter of 3–6 m, depth of 1.5–3 m, and volume of 5–30 m^3. After

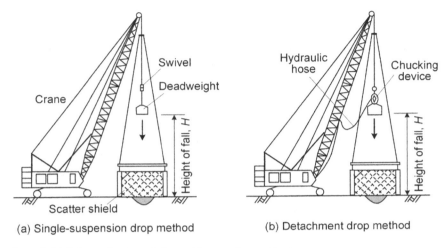

(a) Single-suspension drop method

(b) Detachment drop method

Figure 3.16 Schematic view of the weight tamping method.

tamping, the boreholes are backfilled and levelled by bulldozers in batches in accordance with the crane operation and settlement measurement process. After ground preparation, the next impact phase is implemented. This process is repeated until the prescribed impact energy is applied. A point to be noted in the execution is that the crane that lifts the dead weight must be adequately supported. It is dangerous if the ground surface settles because of fallen dead weight or if the crane loses stability owing to the vibrations. Consequently, the distance between the crane and the dropping point must be sufficient.

The execution efficiency varies considerably depending on the dropping specifications. The execution area per crane and month is approximately 1500–5000 m². Smaller values are for waste ground with an improvement depth of approximately 10 m, and larger values are for rock-crushed ground with an improvement depth of approximately 5 m. However, it should be noted that the efficiency of execution is reduced when improving the ground containing a large amount of fine fraction or on small sites with an execution area of less than 10,000 m². Execution management is characterised by the use of execution adjustment (e.g., tamping control) as necessary, which improves the quality of the improved ground. As the improvement effect varies significantly depending on the soil layer composition and soil condition, it is necessary to conduct test execution, review the design, proceed with execution, or review the improvement specifications through execution management. For quality control, the N value by SPT or cone penetration value by CPT is checked after improvement to confirm that the prescribed degree of compaction has been achieved.

3.3.4 Compaction Grouting

3.3.4.1 Overview

A static compaction method called compaction grouting is also used to compact soft sandy grounds. In this method, a very low-fluidity mortar, fluidised sand, plastic grout, or other improvement materials are forced into the ground to compact the ground. The ground is compacted by the spherical spreading of the mortar or other improvement materials in the ground. This method was developed in USA in the 1950s and was initially used to correct the settlement of settled buildings and fill voids. It was first used as a ground improvement method in the 1970s and has since been widely used to increase strength and prevent liquefaction. As the pipes for injecting mortar and other materials are simply inserted into the ground, and the execution machinery is small, it can be applied for ground improvement directly under existing structures and in narrow places where large ground improvement machines cannot enter.

3.3.4.2 Investigation and Design

The physical properties of the original ground, including its grain size, must be determined as part of the ground investigation required for design. In particular, the fine fraction must be determined as the improvement effect of this method is small if there is a high fine fraction content. It is also necessary to determine the strength of the original ground to set the target ground strength. Specifically, SPTs and CPTs are required. As liquefaction strength is usually evaluated by N values, an SPT should be performed if liquefaction is to be prevented. If the effectiveness of ground improvement is to be assessed by finite element analysis, more detailed investigations, such as undrained cyclic triaxial tests to investigate liquefaction characteristics, are required.

The principle of this method is to reduce the pore space by forcing mortar or other improvement materials into the ground. Therefore, the basic concept of the design is similar to that of the SCP method for sandy soil. First, the required void ratio or relative density after improvement is set and the required improvement ratio (press-in amount) is determined. The improvements in diameter and spacing are then determined according to the required improvement rate. If the SCP method is to be followed, an improvement in diameter, assuming a cylindrical improvement body, is required. In the case of compaction grouting, the improvement material is injected as an irregular bulbous shape in actual execution; however, in design, a cylindrical improvement pile that is uniform in the depth direction is assumed, and the average pile diameter of this pile (referred to as 'converted improvement diameter') is considered to be identical to the improvement diameter for the SCP method. The improvement diameter can be varied freely by changing the injection volume of the improvement material. Therefore, it is necessary to set the converted improvement diameter together with the placement spacing to determine the planar arrangement during design. The standard practice is to use a converted improvement diameter in the range of 0.4–0.7 m and a spacing of 1–2 m. The replacement ratio is 5–15%. By arbitrarily changing the converted improvement diameter, the improvement ratio can also be changed in the depth direction, thereby enabling execution with the optimum improvement rate for each improved soil layer. The so-called 'cut-through execution' method can also be used without injecting the improvement material into soil layers that do not require improvement.

3.3.4.3 Execution

The standard execution procedure for this method is illustrated in Figure 3.17. This method is executed by first drilling a hole to a predetermined depth using a rod with an outer diameter of 73 mm using a boring machine. Next, a device for lifting the injection pipe is installed on the remaining rod, and low-flow improvement material produced by a special injection plant is pumped by a special injection pump and injected from the tip of the rod. After the prescribed injection volume is injected, the rod is extracted by the injection pipe lifting device, and the next injection step is performed. The injection pressure during execution depends on the depth of injection and the hardness of the ground to be improved, but is usually 1–6 MN/m$_2$ and the injection rate is 30–50 ℓ/min. The bottom-up method, in which injection is performed from the deepest point towards

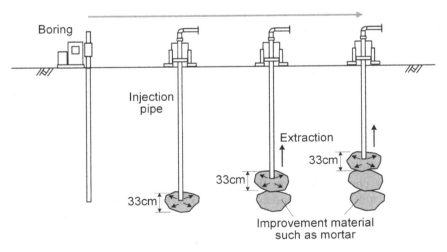

Figure 3.17 Execution procedure of the compaction grouting method.

the ground surface, is the most common method of execution. The injection positions are spaced 0.33 m apart in the depth direction (three injection points per 1 m).

The compactness of the execution system allows it to be easily transported if mounted on a vehicle. If ground improvement is required while an existing facility is in service, the facility can continue to operate during the day, and ground improvement can be performed at night when operations are suspended. For example, at the Tokyo International Airport, this method is used to prevent liquefaction of the ground under runways and taxiways during the night. The method also has advantages such as low vibration and noise owing to compaction grouting using a special injection pump.

One point to be noted is the displacement of the surrounding ground during execution. Although the amount of displacement is smaller than in other compaction methods that use vibratory energy, this is basically a compaction method. Depending on the ground conditions and the depth of the ground to be constructed, the displacement during execution may have an adverse effect on the existing structure. In particular, execution at depths shallower than 5.0 m from the ground level may cause rapid ground uplift. Methods such as prediction using cavity expansion theory have been studied to address these issues, and it is useful to estimate the amount of uplift with reference to the latest research.

For quality control, the N value by SPT or cone penetration value by CPT is checked after improvement to confirm that the prescribed degree of compaction has been achieved.

3.4 CEMENT TREATMENT METHODS

3.4.1 Cement Shallow Mixing

3.4.1.1 Overview

When coastal structures are constructed or ground improvement is implemented, the ground may be overly soft to allow the placement of construction equipment. If the structures are relatively small, only the soil surface layer requires improvement; in this case, shallow ground is treated. Four typical methods are implemented.

1. Spreading method
   This technique involves the sequential spreading of sand and mountain soil. Various spreading methods to produce different layer thicknesses are available; however, on the surface layer of extremely soft ground, sand is embedded to a certain extent, and soft clayey soil is thrusted forward. Consequently, sludge is concentrated in the final execution area. The amount of sludge frequently varies from one place

to another, resulting in unequal settlement. Recently, several attempts have been made to reduce the amount of sinking using lightweight materials for soil spreading.

2. Covering method

   This is a physical treatment method for shallow ground in which the surface layer of extremely soft ground is covered in advance to reduce localised penetration due to sand spreading using materials, such as sheets, rope nets, and bamboo nets. The stresses caused by soil spreading are supported by the tension of the sheet or other materials, reducing the extent of penetration and frequency of penetration fluctuations. When materials with low rigidity are used (e.g., sheets or rope nets), the edges must be firmly fixed. When rigid materials are used, the support provided by the bending resistance is expected, and fixing the ends is less problematic.

3. Cement shallow mixing method

   This treatment approach involves mixing a binder (such as lime or cement) with the soft soil in the shallow layer up to a depth of approximately 3 m. Then, the shallow layer is chemically solidified, such as by pozzolanic reaction. Various types of binders and mixing methods are available. Recently, fixing the edges of sheets as described in Method 2 and partitioning embankments (cement-treated soil panels) have been implemented although these are only used for partial treatment.

4. Drying and drainage method

   If a shallow layer of extremely soft clay soil is allowed to dry naturally, heavy machinery can be set on the untreated site. However, because natural drying requires a long period of time, the combined implementation of forced drainage, culvert drainage, and capillary drying is necessary to shorten the drying period. The available examples of the implementation of the foregoing are relatively few.

Among the four methods, the one presented in 3 directly improves the ground; accordingly, this section describes the shallow mixing method. The covering method described in 2, which uses sheeting, is explained in the section on geosynthetic reinforcement in Chapter 4. In recent years, solidification techniques using lime or cement have been frequently used to increase the stiffness and strength of soft ground. By introducing binders, such as lime or cement, to the ground and then agitating them, soil stiffness and strength can be increased over a short period. When the soil is mixed with lime or cement, the binders react with the soil components, and water in the soil begins to solidify. Virtually all of types of soil can be improved: soft clayey soils and those that are sandy, intermediate, and organic. In the initial stage, the physical properties are improved by ion exchange with clay minerals, followed by considerable hydration reaction. In addition to calcium hydroxide ($Ca(OH)_2$) and calcium silicate hydrate (C–S–H), hydration products, such as ettringite (a needle-like crystalline material), are formed, increasing the soil strength. The medium-term and long-term strengths are further improved by pozzolanic reactions. Strength is affected by various factors and mainly influenced by the water content, soil pH, organic content, cement addition, mixing method, degree of mixing, and physico-chemical and mineralogical properties of the soil. Different from concrete whose quality is relatively constant, the strength of treated soil is characteristically difficult to estimate. Soil that has hardened and gained strength is referred to as treated or stabilised soil; in this book, the term 'treated soil' is used. Lime is used as a binder; however, with the increasing use of cement, the technique is referred to as cement shallow mixing.

The shallow mixing method has the advantages of not generating noise and vibration (typically observed in compaction methods) during the mixing of solidified material. Moreover, the technique only causes minor displacement of neighbouring structures. The method increases the bearing capacity of the shallow layer of soft ground, ensuring the trafficability of equipment for treatment execution. Occasionally, the technique is applied to reclaimed lands near residential areas to prevent the accumulation of water with foul odour, pests, disease-causing organisms, and harmful industrial wastes. However, cement-treated soil is brittle. In particular, soils improved in the form of thin slabs are prone to brittle fractures, such as cracks. Fractures may either be allowed or prevented by implementing design. Different from concrete, cement-treated soils have pore water and do not readily harden; consequently, hexavalent chromium Cr(VI) may leach into the surrounding area. Hence, verifying that the level of contamination does not affect the environment is important.

3.4.1.2 Investigation

The first investigation required in the implementation of the shallow mixing method is an in situ ground investigation to determine formation composition and groundwater level. If the groundwater is in flowing water, the flow velocity must be further investigated. This is necessary because in addition to erosion caused by flowing water, the problem of binder slurry deviation during treatment execution may be encountered. If ground freezing is anticipated, a survey of the freezing depth must be included to implement measures to prevent freezing. In addition to these in situ investigations, laboratory mixing tests for determining the type and amount of binder that must be introduced to obtain the prescribed soil strength must be implemented. Hexavalent chromium leaching tests must also be performed to determine the environmental effects of discharged Cr(VI) at sites with groundwater flow. Physical and chemical tests are conducted to identify the pH level and organic content of soil. These are necessary to identify the strength-increasing characteristics of the soil with the addition of binders.

Figure 3.18 Unconfined compression test of cement-treated soil.

Mechanical tests are performed to determine the properties of the underlying soil to be improved.

A characteristic feature of ground improvement investigations using binders is the laboratory mixing test. In this test, cement or another binder is mixed with the soil collected in situ in varying amounts; then, the mixture strength is measured. Typically, strength is determined by unconfined compression tests, as shown in Figure 3.18. The purpose is to establish the design values (e.g., strength) of the treated soil to be used at the design stage and to determine the type and amount of binder that must be added to achieve a given strength in the field. The trend of strength development is influenced by the type and amount of binder added. Hence, a certain laboratory mix strength is predicted by referring to the soil type and previous mix test data. At least three or more mixes are examined, focusing on the amount of binder addition that can produce the required strength. Several methods for preparing test specimens are available. These include tapping, static compaction, and vibration to fill the mould with pre-solidified soil. The preparation method must be determined by the soil treatment execution method and the physical properties of the soil. However, checking the previous methods used for preparing test specimens is advisable because techniques vary among regions. After preparing the specimens, they are cured under constant temperature and humidity conditions. Then, they are subjected to unconfined compression tests for a specific number of days. The number of curing days must be set according to the conditions; nevertheless, tests are normally conducted for 28 days. Laboratory mixing tests may be excluded if the previous data similar to those of the ground under the same treatment execution conditions can be used. However, the tests must be conducted beforehand as the strength of the treated soil can vary depending on a number of factors. Moreover, the implementation of laboratory mixing tests does not pose a major challenge.

3.4.1.3 Design

The design of the cement shallow mixing method can be divided into two main points: the extent of improvement to achieve the treatment objective and the required strength. The soft soils that can be improved range from reclaimed soils (composed of sludge dredged from sea channels) to alluvial cohesive or highly organic soils that have been consolidated to some extent. The improvement method must be suitable to the soil properties. In soil treatment design, stability is considered from two aspects: external and internal. In the external stability analysis, the extent of improvement is determined, and the sliding and overturning of the treated soil and bearing capacity of the foundation ground are considered assuming that the treated soil has not failed. In the internal stability analysis, the required strength of the treated soil is determined, and the generated compressive, bending, and shear stresses are calculated to ensure that they do not exceed the soil strength.

An example of the design flow for the shallow mixing method is shown in Figure 3.19. The figure shows the design flow for ground improvement with the objective of increasing the bearing capacity. In the design, conditions, such as the ground pressure acting on the ground, are first calculated. Then, the bearing capacity of the original ground is compared with the target bearing capacity; the former must be less than the latter for ground improvement to be necessary. Next, the extent of improvement and required strength are considered. The planned extent of improvement is typically determined by its purpose, and stability is checked by varying the required strength and the ground depth at which improvement is to be implemented. As described, stability (external and internal) is assessed by considering the bearing capacity of the composite ground comprising the improved and unimproved ground beneath. If necessary, the stabilities with respect to punching failure, overall slip, and liquefaction must be studied. In applying the shallow mixing method to extremely soft ground with low shear strength, the extent of improvement and strength of treated soil is typically determined according to flexural strength. Therefore, the use of an analytical method that can simultaneously consider shear stress, flexural stress, and settlement as a design method is advisable. Furthermore, because the strength of treated soil has a limit, determining whether the required strength can be achieved is not consistently possible. The type and amount of binder added must be considered in parallel.

The binder type and amount to be introduced are set based on the results of laboratory mixing tests and the required strength (called design strength) derived from the design studies described above. The procedure is as follows.

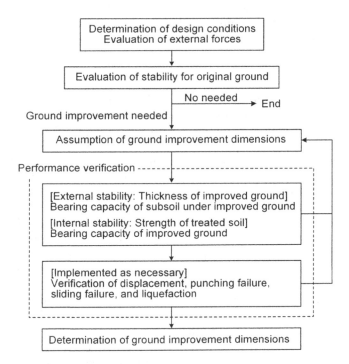

Figure 3.19 Design flow of cement shallow mixing.

First, a binder is selected: the types of binders include ordinary Portland cement, blast furnace cement, lime binders, and quicklime. In Japan, cementitious binders specifically designed to solidify soils that do not facilely harden (e.g., organic soil) and for neutralising hexavalent chromium in soil are commercially available. The selection of a readily available binder based on the properties of the soil to be treated is necessary. Next, the laboratory mixing strength required to achieve the characteristic value of design strength (typically written as q_{uck}; k indicates that this strength is a characteristic value) must be determined. q_{uck} is derived by multiplying the laboratory mixing strength by various coefficients. This is because the field strength is lower and more variable than that obtained by the laboratory mixing test under acceptable mixing conditions. For the variation, the strength is set to 1σ (σ is the standard deviation) or smaller. A coefficient of approximately 0.2–0.8 may also be selected and then multiplied to the actual results. This determines the influence of the execution method considering the difference between the laboratory and in situ strengths. In the case of specific ground with inadequate execution records or large-scale improvement works, the ratio of the in situ strength to laboratory strength can be set after confirming this strength ratio through test execution. Once the laboratory strength is determined, the amount of binder to be added to achieve the target strength can be set based on the laboratory mixing test results.

3.4.1.4 Execution

Two types of execution methods are typically employed for the shallow mixing of soft ground: the stabiliser and backhoe mixing techniques. The stabiliser method is used to improve soils with a depth of up to approximately 1 m. After spreading the binder on the ground surface, the stabiliser agitates and mixes the material on its own, rendering it suitable for improving wide areas. The backhoe mixing method is applied to improve soils at depths of 1–3 m. This technique is suitable for small-scale works and those in which mixing accuracy is not a major concern.

The execution method is as follows. First, the ground is levelled. Then, the machine is set, and a temporary storage place for the binder is located. If the soil to be improved contains grasses, tree roots, debris, or coarse gravel, it must be removed by a backhoe or bulldozer. If water puddles are found in the execution area, drainage measures must be implemented. Next, the area to be improved is divided into small sections. Flexible containers of binders are placed in these areas to regulate the amount of binder to be added. The binder quantity is determined by the design and uniformly spread over the soil to be improved. Spreading is implemented manually, using a machine, or both depending on site conditions, soil conditions, scale of execution, and binder packing. Various spreading methods can be implemented. (1) Binders unpacked from paper bags or (2) discharged from the bottom of a flexible container can be evenly spread. (3) Binders dumped by a truck can be spread using a rake or (4) special spreading equipment. (5) Binders from a lorry can also be evenly spread using a hose. The spreading technique is selected according to the execution conditions. The soil and binder are mixed by stirring, considering the soil properties as well as the type and execution performance of the soil treatment equipment after spreading.

Finally, compaction is performed. Compaction is an important process as it aims to strengthen and stabilise the improved ground, increase its density, and promote chemical solidification. Strength development is minimal immediately after ground improvement (in some cases, strength loss may even occur due to agitation). Hence, the implementation of compaction by running a wet bulldozer or backhoe bucket is normal; however, this must be implemented twice or thrice to avoid over-compaction. The main compaction is implemented 1–3 times after soil improvement. In cases where the water content of the soil is low and rolling is performed while sprinkling water or when rolling cannot be conducted in the same manner as the general implementation owing to site conditions, the amount of binder added may be increased. After the main compaction, curing must be performed such that the soil does not dry until a certain strength is achieved. Heavy machinery traffic must also be avoided until a certain level of strength is achieved.

For execution management, the type and shape of the improver and agitator must be checked before implementing ground improvement. Calibration is also performed to verify the accuracy of the equipment for execution control. During execution, checking the amount of binder added as well as the degree of mixing and depth of improvement is necessary. After execution, workmanship (the extent of improvement) and quality control must be performed. For quality control, cores are obtained to check the strength of the treated soil by unconfined compression tests. The amount of hexavalent chromium eluted is also determined in the laboratory. The characteristics to be observed include foul odour, leaching of hexavalent chromium, and increased pH owing to chemical reactions; these problems are specific to the cement treatment method. During execution, measures must be implemented and monitored to counter the foregoing environmental problems. Countermeasures against foul odours include covering the surface of treated soil after the execution and spraying of deodorants. Special cements may be used as a measure against hexavalent chromium leaching; however, water used for washing must not be allowed to flow outwards.

3.4.2 Deep Mixing

3.4.2.1 Overview

Deep mixing is a widely used ground improvement method in which binders, such as lime and cement, are fed into the ground and agitated. In contrast to the shallow mixing method, which improves a shallow layer of soil to a depth of up to 3 m, deep mixing can increase the depth of the ground improved (hence the description 'deep'). There is a method using a trencher-type mixer, which can improve ground depths ranging 3–13 m. This technique is occasionally referred to as the medium-layer mixing method as it falls between shallow and deep layers. However, because common deep mixing machines can also be used at the aforementioned improvement depths, this book refers to two categories: shallow and deep (Figure 3.20). The principle of improving soft soils is similar to that of the shallow mixing method wherein solidification proceeds via the hydration reaction and hydrate formation of the solidifying material, ion exchange reaction with clay minerals, and pozzolanic reaction (details are presented in the section on the shallow mixing method). In this method, a shaft with agitator blades penetrates the ground. The binder and original soil are agitated and mixed by rotating the agitator blades while discharging the binder. Various shapes of agitation blades have been fabricated, and companies are engaged in the development of new shapes to improve

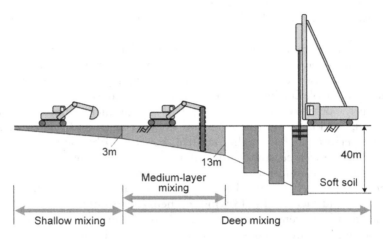

Figure 3.20 Applicable depths of cement treatment methods (land machines).

the uniformity of mixing and cope with mixing hard ground. In addition to the types of equipment with agitator blades attached to the shaft, disc-type and trencher-type agitators can be simultaneously used in the depth direction. However, verifying the characteristics of the execution method being considered for application by examining the literature and actual results is necessary. This is because not all machines can provide the same quality of improvement. The adoption of a low-cost execution method may result in problems, such as the occurrence of points where the design strength is not attained owing to wide variations in strength. When hard ground is to be improved, some machines may be incapable of agitating the soil. In the case of ordinary deep-mixing machines, the scopes for execution are set to N values of approximately 8 and 15 for cohesive and sandy soils, respectively. If harder soils must be improved, the use of a special machine is necessary. Furthermore, the depth of soil that can be improved can vary depending on the machine used. The depth at which improvement can be executed by a large machine is approximately 40 m without connecting work of additional agitator for onshore work. It is 70 m below the water surface (i.e., 50–60 m under the seabed surface) for offshore work.

The deep mixing methods of mechanical stirring described in this section can be classified into the following two categories depending on the condition of the binder to be added.

1. Slurry system
   Cement slurry produced in a plant is electrically or hydraulically pumped to the tip of the agitator shaft of a deep mixing machine. The soft soil and cement slurry are uniformly mixed by the agitator blades

in the area to be improved, creating an improved body with a specified strength.

2. Powder system
   Binders, pneumatically conveyed in powder form instead of slurry, are filled into the ground with an agitator blade and mixed with the soil to create an improved body with a specified strength.

The slurry-based deep mixing method is generally used in port and harbour works where the ground is typically improved underwater. Among the reasons for this is that in powder systems, a considerable amount of air mixes with the binder when it is applied to the ground. The binder with air rises to the seabed surface, causing the seawater to become turbid. Slurry-based deep mixing methods are also increasingly used in land works to ensure the uniformity of the improved body. Jet grouting in which binders are mixed or replace the original soil while cutting the ground with high-pressure water (jet flow) may also be classified as a deep mixing method. This technique is distinctively explained in the next section.

When the method is used in situ, a treated soil column can be formed in a single step. Before the improved soil solidifies, it can overlap and flank to form an improved body of any shape by repeating the process. The most stable form of improvement is the block type, which virtually improves the entire original ground. The grid, wall, and pile improvement types are methods that leave some unimproved sections (Figure 3.21), reducing the

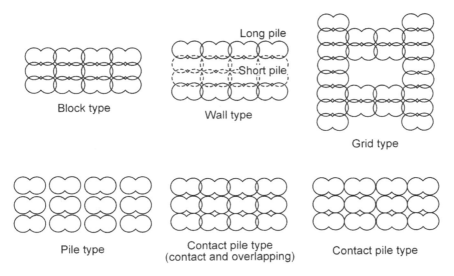

Figure 3.21 Arrangement of cement-treated piles.

expected stability. Nevertheless, these improvement types are economical. The improvement type and ratio are determined by considering the balance between effectiveness and cost. Detailed information on the investigation, design, and execution of the deep mixing method are found in Refs. [9], [10].

The deep mixing method is a ground improvement technology concurrently developed in Scandinavia and Japan in the 1970s. Currently, the technique is used in various parts of the world including Europe, Asia, Oceania, and the USA where its scope of application continues to expand. The execution method itself is fundamentally well established, and the standards and guidelines for design and execution management have been compiled. Ground improvement technology has been developed to the extent that it can enhance soil strength if the appropriate method is implemented; thus far, its application continues to expand. In addition to improving the original ground to support structures, the method has other applications, such as increasing the passive resistance to sheet piles, reducing the main earth pressure on earth retaining walls (such as quay walls), improving the stability of excavation bottoms, protecting underground structures, and preventing liquefaction. More recently, the technology has been employed as barriers to prevent the escape of contaminants from waste disposal sites and as inner embankments to prevent flooding. It has also been used in quay wall-body construction. The typical locations where this technology is used in coastal areas are shown in Figure 3.22. The method's range of application is anticipated to continue expanding. However, several

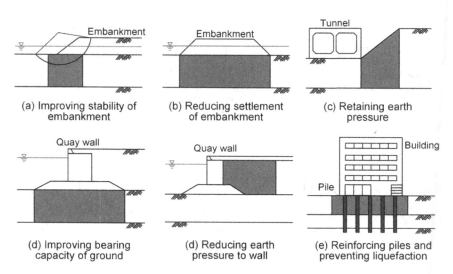

Figure 3.22 Applications of cement deep mixing method.

challenges remain. Global efforts to decrease carbon dioxide emissions to the atmosphere require measures for reducing the use of cement in deep mixing methods. This is because considerable amounts of carbon dioxide are released during cement production. Thus far, endeavours to reduce the amount of cement and other binders used have been insufficient. In situ improvements inevitably result in strength variations, leading to the addition of greater quantities of cement. This is because those implementing improvements may be apprehensive that the prescribed design strength cannot be achieved if low amounts of cement are used. Execution accounts for 70–80% of the cost of the deep mixing method. However, the cost does not increase much although the quantity of cement is increased; consequently, it has been acceptable to increase the amounts of cement. As the history of ground improvement works implemented by deep mixing extends, the re-excavation of previously improved ground becomes necessary. In this regard, the high cost of excavating ground with unnecessarily high strength becomes a problem. In view of the foregoing, methods for reducing the amount of cement added and the use of cement with low carbon dioxide emissions during production must be considered in the future.

3.4.2.2 Investigation

In situ ground investigations are conducted to determine the depth and thickness of the layer to be improved. In addition, laboratory mixing tests are conducted to determine the type and amount of binder to be incorporated such that the required strength can be attained. Hexavalent chromium leaching tests for determining the environmental effects of discharged Cr(VI) are necessary at sites with groundwater flow. Physical and chemical tests of the soil to measure the pH level and organic content must also be conducted to identify its strength-increasing characteristics resulting from the addition of binders. Mechanical tests, such as consolidation and shear tests, are necessary to examine the characteristics of the original ground. This is particularly necessary in cases where certain sections in the vicinity or within the area supposed to be improved are left unimproved. The consolidation test results are used to estimate the final settlement and consolidation time, whereas shear tests are conducted to estimate the shear strength of the improved ground.

A distinctive feature of the deep mixing method is the use of laboratory mixing tests. As explained in the section presenting the shallow mixing method, laboratory mixing tests are conducted by mixing cement in varying amounts with the soil collected in situ and then testing the actual strength. Strength is generally determined using unconfined compression tests. The purpose of the foregoing tests is to establish the design values (e.g., strength) of the treated soil to be used at the design stage. The type and amount of binder to be added to attain the prescribed strength in the field are also determined. However, the strength obtained from laboratory mixing tests does not necessarily reflect the strength developed in the field. Variations in soil properties and differences in mixing methods result in lower soil strength in the field. Because this correlation has been determined from a vast amount of previous data, the strength in the field is estimated according to this ratio. Laboratory mixing tests are recommended but may be omitted if the ground characteristics are well understood based on previous experience in conducting ground improvement at the vicinity of the site. The factors affecting solidification include the soil properties (physical properties, chemical properties, etc.), type and amount of binder added, degree of mixing, and curing conditions. As these factors influence each other, tests must be conducted under specific conditions based on past knowledge. These factors can be summarised as follows.

1. Types of binders: Ordinary Portland cement, blast furnace cement, and lime are used. In Japan, cements specifically formulated to solidify organic soils that are difficult to harden and to inhibit the leaching of hexavalent chromium are commercially available. Blast furnace cement type B is typically used in coastal areas. Blast furnace cement is produced using fine powder of blast furnace slag (a non-metallic molten mineral by-product of blast furnaces in steel mills). The cement is water-cooled and subsequently powdered. It has a dense structure formed by the hydration reaction and excellent resistance to seawater. Blast furnace cement has a low short-term strength; however, in the long term, its strength is high. Further, it has low levels of leached hexavalent chromium and low carbon dioxide emissions during production.

2. Water–cement ratio (W/C): In the case of a slurry system, the W/C ratio of the binder must be varied according to the execution conditions, soil conditions, and target improvement strength. The ratio is typically set in the range 0.6–1.2 although 0.6 and 1.0 are frequently used for marine and land works, respectively. Laboratory mixing tests are generally performed using one value of W/C ratio.

3. Mixing water: The use of water obtained from the site to prepare the specimens is preferred. Nevertheless, if the water from the site is difficult to obtain, tap water may be used.

4. Amount of binder added: The amount of binder added is typically approximately 100–300 kg/m³ of soil to be improved. However, for soils with high organic contents, the binder amount is 200–400 kg/m³ of soil to be treated. Three or four different values are set based on the added binder amounts in the mixing tests.

5. Curing age: The standard curing ages are 1, 2, and 4 weeks; however, 8 and 13 weeks or 1 and 3 days may be implemented. Long-term curing must be considered when slow-setting cement or blast furnace

cement is used or when structures are subjected to design loads after a relatively long period of time.

6. Specimen preparation: Several specimen preparation methods are available. These include preparation by tapping without compaction, static compaction, and dynamic compaction. The recommended method varies according to region, and the suitability of the method depends on the properties of the unsolidified treated soil. The preparation method must be selected based on previous knowledge.

In addition to laboratory mixing tests, hexavalent chromium leaching tests are another feature of the deep mixing method. Different from concrete, cement-treated soil leaches small amounts of hexavalent chromium owing to residual pore water or difficulty in hardening. Tests must be conducted to ensure that leaching does not detrimentally affect the environment at sites with groundwater flow. The binders considered are those containing cement, such as Portland cement, blast furnace cement, special cement, and lime binders. Leaching tests are used to determine the amount of leached Cr(VI) by dipping the treated soil in water. Tests to determine the concentration of Cr(VI) in a suspension are also conducted by breaking up the soil and mixing it with water. The latter method is mandatory in Japan. Specifically, 400–500 g of treated soil is obtained from a specimen (this specimen has been subjected to unconfined compression tests to determine its seven-day strength). The soil is crushed to less than 2 mm, shaken for 6 h, and then subjected to a leaching test. Furthermore, the eluted amount must be less than 0.05 mg/ℓ because this value is the soil environmental standard.

3.4.2.3 Design

The deep mixing design methods for ground improvement vary according to the purpose of the improvement and improvement ratio. When the entire improvement area is solidified as a single unit (block, grid, and wall types), stability is fundamentally considered with regard to two aspects: external and internal stabilities. In the external stability analysis, the extent of improvement is determined. Further, the sliding and overturning of the cement-treated soil and bearing capacity of the foundation ground are considered, assuming that the treated soil is unbroken. In the internal stability analysis, the required strength of the treated soil is determined. The compressive and shear stresses are calculated to ensure that they do not exceed the soil's strength. The design method for pile-type improvement in which the entire improvement area is not integrated has not been fully developed. Presently, the common practice is to focus on compressive strength when designing for vertical load bearing. The improved piles and unimproved area are considered as combined ground, and the shear resistance is evaluated by circular slip analysis when the objective is to achieve slope stability. However, a uniform slip surface has not been observed to develop, and the treated soil piles may either topple over as dominoes or fail in bending when an inclined load is applied to pile-type ground improvement. Accordingly, the expectation is that a design method may be devised. Because civil engineering structures in coastal areas are typically heavy, an improvement type in which the entire improvement area is integrated is preferred.

This section describes the design methods for attaining the two most typical improvement objectives: improvement of the underlying soil supporting the structure and implementation of grid-type improvement for liquefaction prevention. The procedure of the former design method, illustrated in Figure 3.23, is described first. Improvement parameters (extent and type of improvement) have been assumed. As regards the extent of improvement, the area where the overburden load acts is assumed to be the base. If stability is not ensured, the previous assumption regarding the improvement parameters is considered, and a wider improvement area is assumed. However, reaching the bearing stratum may not be possible if a hard intermediate layer that can serve as a bearing stratum to some extent exists in

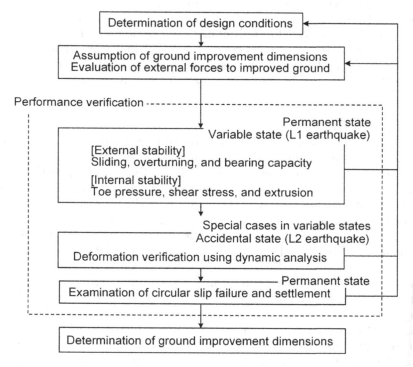

Figure 3.23 Design procedure for cement deep mixing.

the underlying ground, a certain amount of displacement is permitted, or the soft clay layer is extremely thick. The foregoing is called floating-type improvement. As mentioned, different types of improvement (i.e., blocks, grids, walls, and piles) may be implemented. Jointed and unjointed types can also be used depending on the extent to which the improvement piles are lapped. In the design, selecting the appropriate improvement type is necessary. This can be achieved by fully understanding the stability, economic efficiency, workability, and other characteristics of each improvement type according to its purpose, scale, and shape.

Next, external stability is verified. The external force diagrams considered in the design are shown in Figure 3.24. To account for seismic forces, the horizontal seismic intensity must be determined. This intensity is set in accordance with standards and guidelines because different countries and regions have varying ideas on how to determine this value. For example, in Japan, the seismic motion of the engineering bedrock is determined by assuming the epicentre and considering site amplification characteristics to estimate the seismic waveform at the location where the structure is to be constructed. The high-frequency component of this waveform is attenuated to obtain the maximum acceleration on the ground surface using one-dimensional response analysis. This acceleration is then modified according to the earthquake duration to derive the horizontal seismic intensity.

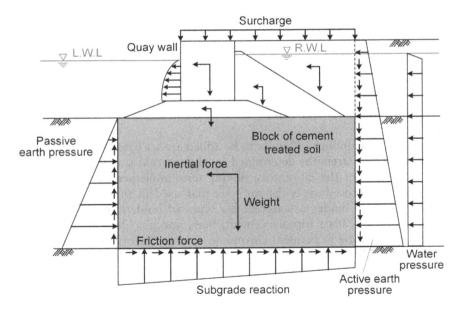

Figure 3.24 External forces acting on cement-treated bodies.

Furthermore, a reduction factor is applied because the improved ground compared with soft soil is less prone to vibration amplification. The horizontal seismic intensity obtained in this manner is used to check whether the improved body does not slide in the horizontal direction, overturn, or sink into the foundation ground (i.e., whether the bearing capacity is secured). For sliding and overturning, verification is performed using equilibrium equations considering the horizontal and rotational directions, respectively. The bearing capacity is verified by checking whether the edge toe pressure at the base of the improved section exceeds the bearing capacity of the foundation ground. Standards and guidelines are used as reference for the partial coefficients of each mode. If the external stability is not satisfied, the previous assumption regarding the improvement parameters is considered. Moreover, the improvement width is increased, and the examination is repeated until the stability is satisfied.

The internal stability of ground improvement is verified after checking the external stability. The stability is confirmed based on the generated ground stress so that shear failure of the improvement or edge toe failure of the section where forces are concentrated does not occur. The stress is calculated by considering the improved body as elastic. In the case of wall-type improvement, checking whether the cohesive soil between the improvement walls does not extrude is necessary. In the calculations, the external forces, including seismic forces, shown in Figure 3.24, are applied to the improvement. In the case of complex structures or improvement types, stresses may be determined by finite element analysis method. The required strength of the treated soil is set equal to the calculated stress. However, determining the required strength of the treated soil is not invariably possible because a strength limit exists. The type and amount of binder added must be concurrently considered. In previous experiences, the standard design strengths used by typical machines for executing ground improvement are q_{uck} = 800–2500 kN/m² for marine works on clayey soils, q_{uck} = 100–1000 kN/m² for land works on clayey soils, and q_{uck} = 500–1500 kN/m² for land works on sandy soils.

The type and amount of binder to be added are set based on the required strength (design strength) determined by the internal stability verification and the results of the laboratory mixing tests implemented in the investigation. This procedure is the same as that used in the shallow mixing method. First, a binder is selected; the types of binders include ordinary Portland cement, blast furnace cement, lime, and quicklime. Cement capable of neutralising hexavalent chromium can also be obtained. The selection of a readily available binder based on the properties of the soil to be treated is necessary. Next, the laboratory strength that reflects the design strength (characteristic value), q_{uck}, must be determined. The design strength is obtained by multiplying the laboratory strength by various coefficients, as given by the following formula:

$$q_{uck} = 0.67 \cdot \overline{q_{uf}} = 0.67 \cdot \lambda \cdot \overline{q_{ul}}, \tag{3.21}$$

where q_{uck} is the characteristic value of the design strength; q_{uf} is the unconfined compressive strength in situ; q_{ul} is the unconfined compressive strength in a laboratory; and λ is the in situ–laboratory unconfined compressive strength ratio. The coefficients are multiplied as shown in the foregoing formula because the in situ strength is lower than that obtained by the laboratory mixing tests conducted under ideal mixing conditions. The coefficients account for the variations in the treated soil, execution method, and other factors. For the variations in the treated soil, 0.67 is multiplied to reduce the strength by 1σ (σ is the standard deviation). For the execution method and other factors, the coefficients of λ (0.5–1.0 (small execution vessels), 0.8–1.0 (medium and large execution vessels), 0.5–1.0 (land execution machines)) based on actual results are multiplied. In the case of using special soils or binders for which experience in ground improvement implementation is limited, the in situ–laboratory strength ratio can be set after confirming its value through test execution. Once the laboratory strength is determined, the amount of binder to be added to achieve the target strength can be determined from the results of the laboratory mixing tests. However, a minimum binder amount must be added to ensure the uniform quality of the improvement in the field; hence, this quantity must also be considered.

To facilitate comprehension, the term 'design strength' has been used in the foregoing. However, the strengths actually used in the calculations for verifying the internal stability are design compressive, shear, and tensile strengths; the magnitudes of these strengths are derived from the design strength. The design compressive strength, f_{ck}, is obtained by multiplying the design strength, q_{uck}, by $\alpha\beta= 0.8$. This factor accounts for the uncertainties in the strength of the overlap between the improved piles and partially remaining unimproved sections. The design shear strength, f_{shk}, is half of the design compressive strength, f_{ck}, and the design tensile strength, f_{tk}, is derived by multiplying the design compressive strength, f_{ck}, by 0.15. The maximum design tensile strength, f_{tk}, is 200 kN/m².

When the deep mixing method is used to prevent liquefaction, a grid-type improvement is often used. Liquefaction prevention using grid-type improvements involves casting high-strength walls in the underlying soil in a grid pattern. This restrains the shear deformation of the underlying soil in the grids, thereby controlling the increase in excess pore water pressure. Grid-type improvement is more economical than block-type improvement and has been increasingly applied; its effectiveness has been confirmed in actual earthquakes. The external and internal stabilities of the design method must be considered in the same manner as stability. However, the problem is the liquefaction countermeasure effect, which is set by grid spacing. The excess pore pressure generated can be reduced by decreasing the grid spacing, L, in the vibration direction. Dividing the grid spacing by the thickness of the liquefied layer, H, such that L/H is generally less than 0.8 is considered appropriate. However, this method results in an excessively wide grid spacing particularly when the liquefaction layer is thick. Moreover, the grid spacing is unreasonably narrow when the layer is thin. For this reason, the author proposes dividing the grid spacing, L, by the depth, d, at which liquefaction is to be suppressed; this sets L/d to 1.4 or less [11].

Although this book presents examples of design methods for improving underlying soil-supporting structures and preventing liquefaction, deep mixing methods have also been applied to other areas. As mentioned, the deep mixing method is also used to increase the passive resistance of sheet piles, reduce the main pressure on earth retaining structures (such as quays), improve the stability of excavated bedrock, protect underground structures, and improve embankments to prevent flooding. The basic concept for these designs remains the same: the external forces acting on the improvement must be considered, and the external and internal stabilities must be checked. Moreover, additional items, such as liquefaction control effects in grid-type improvements, must be considered for each application.

3.4.2.4 Execution

The general executions for the deep mixing method are illustrated in Figure 3.25. The two most common methods for discharging the binder are penetration and withdrawal discharges. Each method of binder discharge has its own advantages and disadvantages. Penetration discharge is

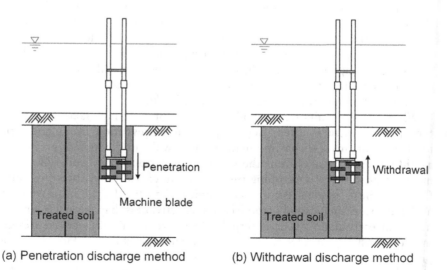

(a) Penetration discharge method   (b) Withdrawal discharge method

Figure 3.25 Execution of cement deep mixing.

advantageous in terms of jointing and stir mixing the improvement pile. However, if the execution machine encounters a problem, the solidified binder may obstruct the subsequent penetration or withdrawal. Conversely, withdrawal discharge is disadvantageous in terms of the jointing and agitation mixing of the improved pile; nevertheless, the foregoing does not compromise safety.

The execution and management of the slurry system deep mixing method used in coastal areas are described in this section. A diagram of the execution vessel used for marine works is shown in Figure 3.26. The vessel has multiple agitator shafts (2, 4, and 8) and is classified according to the area that can be improved by single penetration and withdrawal: large, medium, and small vessels (approximately 5.7, 4.6, and 2.2 m^2, respectively). As a standard, the improvement for large and medium vessels is implemented by withdrawal discharge. Small vessel improvement is implemented by penetration discharge. The quality control items for marine works include the amount of binder, degree of agitation and mixing, placement position, verticality of the agitation axis, placement depth, and bottoming. In particular, the degree of agitation and mixing of cement into the ground (the most important factor) is conventionally controlled by the rotational speed and the penetration/withdrawal speed of agitator blades, which are difficult to directly monitor during execution. Although not as severe as the compaction method, the deep mixing method also raises the original ground during execution because cement slurry is injected. The shape of the upheaval greatly varies depending on conditions, such as soil property, thickness of the improved layer, improvement ratio, execution sequence, thickness of the upper unimproved layer, and presence or absence of existing structures in the vicinity. In most cases, the upheaval is approximately 0.7 times the amount of cement slurry injected. Cement may be mixed with the upheaval; however, the resulting soil quality is variable. Cores are typically extracted after improvement, and the strength of the treated soil is checked using unconfined compression tests.

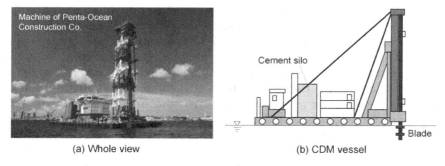

(a) Whole view   (b) CDM vessel

Figure 3.26 Marine machine for cement deep mixing. (Courtesy of Penta-Ocean Construction Co.)

For land works, the machinery for improvement execution consists of a treatment machine and slurry plant. The standard number of agitator shafts is 2, 3, or 4. The slurry is generally discharged by penetration discharge. Withdrawal discharge is occasionally used when a hard stratum in the soil layer to be improved exists or when the execution is implemented at considerable depths. The execution management and quality control for land works are similar to those for marine works. In many cases, the volume of upheaval is 70–80% of that of the injected cement slurry. A particular problem in land works is the lateral displacement that occurs in the surrounding ground during execution. Although the displacement is small at a distance equivalent to the depth of improvement, several tens of centimetres of displacement may occur in the immediate vicinity. Hence, accounting for displacement in the design is necessary.

3.4.3 Jet Grouting

3.4.3.1 Overview

When previously constructed civil engineering structures are reinforced or new ones that are in close proximity to existing structures are built, the implementation of ground improvement without demolishing the existing structures is occasionally preferred. The more developed and mature the social infrastructure, the more the opportunities for retaining existing structures. This necessity is also increasing in coastal areas. The ground improvement methods introduced in the previous sections (with the exception of compaction grouting) are typically applied to newly constructed civil engineering structures. These methods are not easy to implement in the vicinity of existing structures. This section introduces jet grouting, which can be used to improve ground near existing structures. Jet grouting improves the soil by cutting and breaking the ground with considerable power generated by applying high pressure to the fluid and agitating and mixing the binder and soil. The improvement principle is based on the hydration reaction between the binder and pore water of soft soil, ion exchange between the hydration products and clay minerals, chemical reactions (mainly pozzolanic reaction), and soft soil replacement (full or semi-replacement) by a soil–binder mixture. Different from the deep mixing method, which involves mechanical stirring, the soil is replaced by a soil–binder mixture that eventually hardens into high-strength ground. Another advantage of jet grouting is that if the structure can withstand high-pressure jets, the ground in close contact with the structure can be improved. This method was developed in Japan in the 1970s and subsequently used in Italy to repair building foundations. Since then, its use has been popularised throughout Europe, Japan, and the rest of the world. Recently, a combined machine that enables the deep mixing of the inner part of the treated soil pile by mechanical agitation and the outer part by high-pressure jetting has been developed.

Jet grouting for ground improvement can be used directly under existing structures and in narrow areas where large ground improvement machines cannot enter. This is because the rod is simply inserted into the ground, and the execution equipment is small. However, airlift is required to transport the cuttings to the surface. Moreover, the implementation slime treatments are necessary for binder circulation. The jet grouting technique has been applied to various locations. Its use in coastal areas includes water sealing around sheet piles, increasing the passive resistance in front of sheet piles, reinforcing revetments and bridge foundations, and implementing countermeasures against subsidence and liquefaction in areas where existing structures are located, as shown in Figure 3.27. Depending on the machine used for jet grouting, improving clay and sandy soil with N values of up to 10 and exceeding 50, respectively, is possible. Some machines are also capable of improving soil at depths of up to 60 m. Jet grouting is relatively expensive but highly versatile. In recent years, the cost of execution has been reduced by increasing the improvement volume and speed of execution.

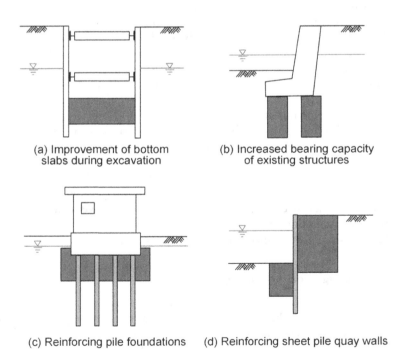

(a) Improvement of bottom   (b) Increased bearing capacity
slabs during excavation   of existing structures

(c) Reinforcing pile foundations   (d) Reinforcing sheet pile quay walls

Figure 3.27 Application of jet grouting.

Further information on jet grouting is found in Ref. [12], which provides an excellent summary of the method.

Various types of jet grouting methods can be classified according to the number of tubes used; the methods can be broadly classified into the following four categories.

1. One-phase flow system (single-tube method)
   This method produces a columnar solidified body by rotating and pulling up the borehole rod while injecting the binder through a single-hole nozzle attached to the sidewall of the rod. This is the oldest jet grouting technique. Because the jet energy is attenuated by groundwater, the treated soil piles produced are smaller in diameter than those produced by other methods that use air. It also has some disadvantages, such as its ability to induce ground uplift. However, this technique continues to be used because the equipment employed is smaller than that of the other methods.

2. Two-phase flow system (double-tube method)
   This method reduces the attenuation of ultra-high-pressure fluids owing to the influence of groundwater by injecting the binder with air. This is also known as the double-tube method because double-tube rods are employed to secure the binder and air paths.

3. Three-phase flow system (triple-tube method)
   This method involves injecting air and water and separately injecting the binder. It is also known as the triple-tube method because triple-tube rods are used to ensure that three paths are provided for water, binder, and air.

4. Perforated pipe method
   The perforated pipe method uses rods with multiple holes (fluid pathways) that are sparsely spaced; they differ from rods with a concentric circular cross-section. The diameter of a pile for ground improvement can be increased by applying jetting through multiple holes. A pile with a uniform diameter can be driven using intersecting multiple jets. This approach is commonly used to achieve improvement by oscillation. By oscillating the rod, fan-shaped bodies instead of circular solidified ones are also formed, facilitating the production of any shape.

3.4.3.2 Investigation and Design

In addition to in situ ground investigations for determining the depth and thickness of the layer to be improved, geophysical tests are required to identify the characteristics of the original ground. In particular, when the ground is sand or gravel, investigating the grain size and permeability coefficient is necessary. In the case of humic and organic soils, the organic matter content should also be determined because the jet may not be able to cut through if the soil contains fibrous materials. Different from deep mixing, which involves mechanical stirring, jet grouting does not involve laboratory mixing tests except for the single-tube method. The main reason for this is that the amount of water and binder added and mixed with soil to be improved cannot be clearly defined. The amounts of binder injected and mixed are not necessarily equal. They depend on factors including the binder used and waste mud volume and concentration. Although hexavalent chromium leaching tests are implemented, unambiguously determining the mixture of soil and binder in specimens is extremely difficult. A soil–binder ratio of 1:1 by volume is used as a reference.

Jet grouting can also be used in combination with improvement piles to form a block (lap or contact arrangement), grid, wall, or pile type improvement. In the block-type lap arrangement, the piles used for improvement are overlapped to provide full surface improvement and seal water inside the improved area. If watertightness is not required, a contact arrangement is used. Once the solidified body is formed, the subsequent stability of the structure is verified using the deep mixing method of mechanical stirring. The high-pressure jet stirring method differs from the deep mixing method of mechanical stirring in terms of the strength of the solidified body and diameter of the pile used for improvement. These points are explained below.

The strength of the treated soil depends on the binder used. However, each method specifies the binder and the amount to be added such that the design strength values are approximately 3000 and 1000 kN/m^2 when the original ground is sandy and clayey, respectively. However, note that based on experience, these values represent minimum strengths, and the actual strength of the treated soil is greater. The binder used in jet grouting must enable the cementitious particles to be pumped as a fluid slurry and discharged smoothly from the injection nozzle. For this reason, the binder used for each method is specified. However, cementitious binders with an admixture for increasing fluidity are commonly used, and their formulations are specified. The binders and admixtures used are generally approved for each execution method and may be obtained in pre-mix form.

In jet grouting, the execution specifications (e.g., jet pressure, jet flow rate, and lifting speed) are defined for each method, and general values of the effective diameter for design are given. The piles with diameters in the range 1.0–2.0 m are expected to be strengthened by a solidified columnar body. These diameter values vary from one execution method to another and are found in technical manuals. In certain methods, the diameter may be controlled to a certain value. The effective diameter for design is typically set based on the grain size of the soil to be improved, N value, and improvement depth. In the case of cohesive soils, the cohesive strength is generally considered in addition to the N value. This is because highly cohesive soils significantly reduce the cutting ability of the high-pressure fluid, resulting in a smaller improvement diameter. In principle, if soil layers with different soil types (e.g., clayey and sandy soils) are present within the improvement area, the soil type with the least strength and the soil layer with the smallest improvement diameter after treatment must be considered.

3.4.3.3 Execution

Jet grouting is implemented after completing prior tasks, such as preparation and plant assembly. The implementation process differs in some areas depending on the execution method, but the fundamental procedure involves drilling, rod insertion, and jetting. The drilling machine is installed at a predetermined position, and the hole is drilled to a predetermined depth. Two types of drilling methods may be used: direct drilling using rods (that inject ultra-high-pressure fluid) and drilling using casing pipes. Next, the rod is inserted, and the casing pipe is withdrawn. If the borehole wall collapses, some casings may be left after failure. Finally, jetting is performed. A high-pressure pump is used to inject the fluid at a predetermined flow rate and pressure while the rods rotate. Then, the rotational speed of the rods is increased to a predetermined speed.

Machinery can be broadly classified into equipment for drilling and execution. The drilling equipment consists of drilling machines (boring machines), pumps, and tools. Borehole drilling machines are generally of the rotary type (rotary boring); they used for drilling injection holes and installing forming rods. The borehole diameter is determined by the drilling method and ranges 40–250 mm. The execution equipment consists of a jetting machine, ultra-high-pressure pump, cement slurry plant, grout pump, and injection equipment; it can also be combined with a drilling machine. A rafter crane is generally positioned at the execution point to move the machine. Ultra-high-pressure pumps must satisfy the execution specifications for each execution method. Two pumps may be used depending on the execution method. Cement slurry plants are used to mix and blend various hardening materials. If these facilities are vehicle-mounted, they can be moved easily. If ground improvement is to be implemented while an existing facility is in service, the facility can continue to operate during the day and ground improvement can be implemented at night when the facility is temporarily out of service.

Another point to consider is waste mud disposal. Because mud disposal significantly impacts execution costs, various methods have been proposed to reduce the amount of waste mud. In general, the following two methods are used for mud treatment.

1. Direct discharge method
   The material is collected in a ditch at the execution location (if this is not possible, a tank is installed), fed into a vacuum truck or tank truck using a sand pump, and transported to a designated disposal site. For

transport and disposal, the services of a contractor licensed to dispose industrial waste are required; this is in accordance with the provisions of the Waste Disposal and Public Cleansing Act.

2. Temporary storage system
   The discharged waste mud is placed in tanks where it solidifies in approximately one day owing to its cement content. It is removed using equipment, such as hydraulic excavators and discharged by dump trucks. This approach requires a minimum of two days' worth of storage in tanks that must be available in the yard. Solidified waste mud remains an industrial waste product and must be appropriately disposed.

Execution management involves measuring the rod penetration, jet discharge, jet discharge pressure, and rod discharge angle of the execution machine. This management regulates machine operations without encountering any problems. To determine whether the implemented improvement is effective, rods with attached cameras are inserted into the ground. Sound waves and temperatures are also measured to ascertain whether the jets reach the target improvement site. For quality control after execution, soil cores are typically extracted to check the strength of the treated soil through unconfined compression tests. Recently, simple quality control methods, such as PS logging, have also been studied.

3.5 CHEMICAL GROUTING METHOD

3.5.1 Overview

Compaction grouting and jet grouting have been introduced as ground improvement methods that can be used in the presence of existing structures. Another representative method is chemical grouting. Chemical grouting is a method for modifying soil properties (such as soil strength) by injecting a 'chemical' liquid into the ground; the liquid hardens over time under pressure. Compared with other ground improvement methods, chemical grouting is characterised by the use of compact execution machinery and easy mobility. Accordingly, it is widely used to increase the pinpoint strength in narrow construction areas and directly improve the ground under existing structures. This ground improvement method has the least impact on existing structures because it only involves chemical injection without exerting the same force used in jet or compaction grouting. Chemical grouting is relatively expensive, but the cost has been reduced in recent years.

The methods are mainly categorised into two: fracture grouting (in which the injected suspension-type chemical breaks up the structure of soil particles as it enters) and permeation grouting (in which the solution-type

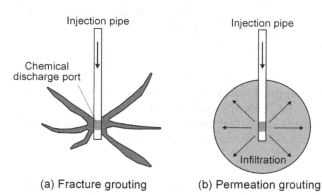

Figure 3.28 Fracture and permeation grouting.

chemical penetrates pore spaces without breaking up the structure of the soil particles) (Figure 3.28). The latter method is used for ground improvements in coastal areas and involves the use of a permanent solution-type chemical. This approach is purported to have started in the 1800s. At that time, clay–lime mixture slurry was reportedly pumped into the foundations of a scoured stone revetment in a French harbour. This marks the beginning of chemical grouting with a history of more than 200 years. In the 1900s, two types of injection materials—suspension and solution—were used in Germany, Belgium, and other countries; hence, chemical grouting was developed. Originally, chemical grouting was commonly used for temporary works on land. In the 1990s, solution-type permanent chemicals were used in Japan as a countermeasure against soil liquefaction. Currently, these chemicals are widely used for ground improvements in coastal areas. Solution-type chemicals are active silica-based chemicals purified from water glass that does not contain suspended material. These chemicals can uniformly penetrate sandy substrates and permanently solidify them. Polymer-based (acrylamide, urea, and urethane) chemicals had been previously applied; however, their use has been discontinued because these chemicals contaminated the ground. The chemicals are classified according to their hardening time (gelling time), that is, from instant (a few seconds) to slow hardening. The hardening time is adjusted by modifying the mixing method of the main material and admixture. In particular, the use of a solution-type slow hardening chemical with an appropriate solidifying time to modify the physical properties of sand such that the chemical becomes a permeation injection is advisable. Although the penetration time is longer if the soil contains fine fraction, modifying the physical properties of sand containing some fine particles to become homogeneous by delaying the hardening time is also possible.

The main purposes of ground improvement in coastal areas using solution-type chemical grouting are liquefaction prevention, water interception, and prevention of soil suction. Liquefaction control is typically implemented in the vicinity of underground structures or directly under runways and taxiways of maritime airports where ground displacement is a problem and compaction methods cannot be employed. The implementation of improvements directly under existing structures is also possible. Bend drilling using gyrosensors enables the injection of chemicals at arbitrary locations on the ground. For example, in the liquefaction countermeasures at Tokyo International Airport (described in the section on compaction grouting), holes for compaction grouting are drilled on runways and taxiways at night. In contrast, bent boreholes for chemical grouting are drilled at the sides of runways and taxiways. The chemicals are injected while the facilities are in use. In coastal areas, chemical grouting is necessary to intercept water and prevent suction. The ground behind the seawalls is sucked out by wave action, which causes problems, such as ground sinking at the rear of the seawall. Pinpoint chemical grouting can be used to prevent water runoff.

3.5.2 Investigation

The required investigations for chemical grouting design include the in situ ground investigations of the depth and thickness of the layer to be improved and a survey of the surrounding groundwater. Based on the results of this ground investigation, the suitability of the method is examined to determine whether the improvement is effective. Legal concerns and execution conditions, such as the size of the work area, time of day when the execution can be implemented, vibration, and noise, must also be checked. Physical tests are necessary to investigate the characteristics of the original ground. In particular, because this technique is a ground improvement method in which chemicals permeate the ground, particle size distribution is an important aspect. If the fine fraction content is high, ground fracture occurs and homogeneous improvement is not possible. In sandy soils with a high coefficient of uniformity, solution-type chemicals are easily affected by dilution, and the strength of the solidified body tends to be low. These effects are discussed below. Furthermore, when solution-type chemicals are applied, pH tests of the soil suspension and mixing tests using chemicals and local soil must be conducted. Silica-based solution-type chemicals have a viscosity similar to that of water and a specific gravity that is 6–8% heavier than that of water. Moreover, the improvement effect after execution is greatly influenced by the particle size distribution of the target soil.

Because the grain size distribution is an important aspect of chemical grouting, its influence must be described. For the fine fraction, F_c, the normal injection speed is in the range 10–15 ℓ/min. For sands with F_c less than 25%, the injection type is mainly permeation grouting; when F_c is exceeds 25%, the injection type gradually shifts to fracture grouting. When $F_c > 40\%$, the injection type is virtually fracture grouting. If the silt content is non-plastic, F_c is high in the transition zone. The uniformity factor, U_c, is the ratio of 60% to 10% of the grains passing through a sieve. Sandy soils with uniformity factors of up to approximately 2 predominantly have one grain size. They are less compact and have large voids, rendering them suitable for the permeation of solution-type chemicals. Sandy soils with a large U_c value have a wide grain size range. They are easy to compact and have relatively small voids. For soils whose particles have large effective diameters, dilution has a significant effect. Spherical compacted bodies are less likely to be obtained, and the compaction strength is low. For an average particle size (D_{50}) exceeding 0.5 mm, even if the pore space is filled with solution-type chemicals, the bonding force among the particles remains low owing to the large pore space; the solidification strength is also low. Note that the pore spaces are replaced by a gel-like substance; therefore, liquefaction is less likely to occur. However, the effect of increased shear forces, such as that of reduced seismic earth pressure, is small.

3.5.3 Design

Chemical grouting is a type of ground improvement method that solidifies the ground; however, the strength of the improved ground, especially in solution-type chemical grouting, is lower than that achieved by cement treatment methods. The unconfined compressive strength is approximately 50–100 kN/m². Therefore, adopting design techniques, such as replacement, consolidation-accelerated, and compaction methods, that improve ground performance is typical. This is more common than considering the improved soil as a solidified body that behaves as a single unit. Therefore, a minimum extent of improvement is assumed, and if the performance is not satisfactory, the degree of improvement is extended. When chemical grouting is used to prevent liquefaction, most of the pore water is replaced by a solidified gel-like chemical. Consequently, excess pore water pressure does not propagate into the improvement area as it does in the compaction method; hence, improving an extra area is unnecessary.

The shear and liquefaction strengths of improved soil are considerably influenced by injection specifications. Six injection specifications are required for execution: chemical type and concentration, injection rate, injection speed, injection pressure, injection hole spacing, and hardening time. These are described as follows.

1. Type and concentration of chemicals
   The chemical type must be selected to satisfy the required improvement specifications. The selection criteria include permeability and strength as well as the permeability required after the improvement.

Subsequently, mixing tests are implemented to determine the appropriate mix.

2. Injection rate

   Injection rate is the multiplying factor for the void ratio of the ground and the filling rate of the chemical into the pores. For each method, the standard void ratio, filling rate, and injection rate are set based on the N value from the SPT. For silty sand containing fine fraction, the injection rate must be set using the injection model tests.

3. Injection speed

   The injection speed is typically determined through a critical test using water. However, in the stage of design, the injection speed is generally set based on the speed used in previous execution projects. Note that if restrictions exist at the time of a single injection application, the amount of water injected per day is limited by the application time and injection speed. The quality of the treated soil tends to deteriorate when the injection speed is high because fractures easily appear on the ground. The more gradual the injection speed, the better the quality of the improved ground.

4. Injection pressure

   When improving pavements or the ground directly under buildings, the upper limit of injection pressure must be set. Although this pressure limit varies according to the execution method, execution management is typically introduced when an injection pressure upper limit is set. The injection speed is reduced to a constant rate if the injection pressure exceeds the upper limit. The upper limit of the injection pressure differs for each method and must be set with reference to the corresponding technical manual of the execution method.

5. Injection hole spacing

   The spacing among injection holes represents the extent of improvement assigned to one injection pipe. Therefore, the injection hole spacing can be determined based on the relationship between injection volume and volume assigned. Because a wide injection hole spacing increases the injection time and renders the control of the hardening time more difficult, an upper limit for the hole spacing may be set.

6. Hardening time

   The hardening time is called the gelling time and refers to the time for the chemical solution to react and solidify. For special silica-based solution-type chemicals, the gelling time is controlled by the pH level as it varies with the pH level of the chemical. The time that elapses from the moment the chemical is injected into the ground to the instance when the soil solidifies is called the soil gelling time. For quality control, setting and regulating the injection and soil gelling times such that they are virtually simultaneous are important. When the in-soil gelling and injection times are fundamentally the same, the quality of the improvement is high because it is not affected by dilution. If the rate of in-soil gelling is slower than the injection rate, the solution-type chemical with low viscosity and high specific gravity solidifies as it drips; the strength is also low. Conversely, if the in-soil gelling time is shorter than the injection time, the extent of solidification is hetero-geneous because the ground fractures; however, the strength is high because it is not affected by dilution. These characteristics must be considered when setting the gelling time.

3.5.4 Execution

In the case of large-scale work or when chemical grouting is applied for the first time to a location, conducting field injection tests before implementing the main work to confirm whether the improvement specifications satisfy the target is advisable. If the test work is difficult to implement, the conduct of injection model tests using soil tanks filled with local sand to confirm the shape of the improved ground is also effective. Two types of chemical grouting methods may be used: single-tube and double-tube grouting. The double-tube bag-packer method is commonly applied to allow solution-type chemicals to infiltrate into sandy ground for counteracting liquefaction and other purposes. In this method, bags are installed above and below the chem-ical discharge port, as shown in Figure 3.29. The bags are inflated before injection to seal the borehole walls by allowing them to adhere closely to the walls.

The execution management for the injection of solution-type chemicals is described herein. The viscosity of the chemical solution is as low as

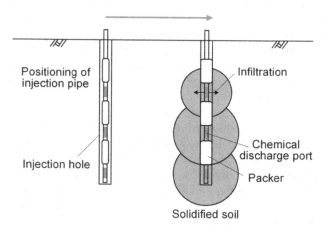

Figure 3.29 Chemical grouting using bag packers.

that of water, and its specific gravity is 6–8% higher than that of water. Accordingly, the following points must be considered to improve ground quality after treatment.

1. Mixes must be set such that the soil gelling and injection times are approximately the same in the chemical mix tests. Quality control must be implemented during execution by measuring specific gravity and pH before work starts.
2. The injection hole spacing and speed must be determined based on past experiences. However, the treatment quality can improve if the injection speed is reduced to the extent permitted by the execution conditions.
3. If the area to be improved has alternating layers with widely differing permeabilities, the use of cement bentonite or similar materials in combination with the filling is effective.
4. If significant leakage or ground surface uplift occurs, reducing the number of injection points and the amount of chemical injected is effective.
5. When restrictions on the upper structure displacement exist, an execution method in which the displacement is measured with a control reference value is used. The injection speed and pressure are reduced such that the displacement does not exceed the reference value.

For quality control, borings and soundings are performed after the execution. For more accurate quality control, undisturbed samples of the solidified soil are extracted by boring. Then, unconfined, tri-triaxial, and cyclic triaxial compression tests are performed. Block sampling is ideal; however, in cases where this is difficult to implement, triple-tube sampling must be performed. Thin-wall sampling may disturb the specimen resulting in low strength estimates. If undisturbed samples are difficult to obtain even by triple-tube sampling, higher quality sampling, for example, by using gels, is required. Quality control may be implemented by sounding. However, the unconfined compressive strength of the ground improved by solution-type chemical grouting is approximately 50–100 kN/m^2. Its difference from the strength of the unimproved area is typically small; hence, determining whether the ground strength has indeed improved is difficult. Future improvements in measurement accuracy are expected.

References

1. Ports and Harbours Bureau, Ministry of Land, Infrastructure, Transport and Tourism (MLIT). 2018. *Technical Standards and Commentaries for Port and Harbour Facilities*. The Ports & Harbours Association of Japan (in Japanese).

2. Barron, R.A. 1948. Consolidation of fine-grained soils by drain wells by drain wells. *Transactions of the American Society of Civil Engineers* 113(1): 718–742.
3. Yoshikuni, H. and Nakanodo, H. 1974. Consolidation of soils by vertical drain wells with finite permeability. *Soils and Foundations* 14(2): 35–46.
4. Hansbo, S. 1981. Consolidation of fine-grained soils by prefabricated drains. Proceedings of 10th International Conference on Soil Mechanics and Foundation Engineering, 677–682.
5. Kitazume, M. 2005. *The Sand Compaction Pile Method*. CRC Press.
6. Takahashi, H., Kitazume, M., and Maruyama, K. 2009. Deformation behaviour of SCP improved ground to limit state. Proceedings of the 17th International Conference on Soil Mechanics and Geotechnical Engineering, 2272–2275.
7. Kirsch, K. and Bell, A. 2013. *Ground Improvement*. CRC Press.
8. Kirsch, K. and Kirsch, F. 2017. *Ground Improvement by Deep Vibratory Methods*. CRC Press.
9. Kitazume, M. and Terashi, M. 2013. *The Deep Mixing Method*. CRC Press.
10. Kitazume, M. 2022. *Quality Control and Assurance of the Deep Mixing Method*. CRC Press.
11. Takahashi, H., Kitazume, M, and Ishibashi, S. 2006. Effect of deep mixing wall spacing on liquefaction mitigation. Proceedings of the International Conference on Physical Modelling in Geotechnics (ICPMG), 585–590.
12. Croce, P., Flora, A., and Modoni, G. 2014. *Jet Grouting: Technology, Design and Control*. CRC Press.

Chapter 4

Use of Improved Soil

This chapter describes ground improvement methods that effectively use soil that has been improved at a site different from the in situ location and moved to an area where a civil structure is to be constructed. This corresponds to a type of ground improvement technique to enhance the ground such that it can be used for civil engineering structures. The following methods are specifically introduced: mixing soil with cement and other solidifying materials, followed by site placement before the mixture hardens; recycling and using by-products as solidifiers or ground materials; mixing cement and soil in pipes to produce cement-treated soil; mixing foam to produce lightweight cement-treated soil; filter pressing to separate soft soil from water; using solidified soil produced by the deep mixing method; modifying the properties of the entire ground at a site by utilising geosynthetics.

4.1 PRE-MIXING METHOD

4.1.1 Overview

In the shallow and deep mixing methods, cement is added to the soil in situ to solidify the ground. In contrast, in the pre-mixing method, soil and solidifiers are blended in a plant mixer. Then, before they solidify, they are placed at a predetermined location where a structure is to be built. In this method, solidifiers and anti-separation agents are mainly added to sandy soil, mixed, and then transported in an unsolidified state. The mixture is then backfilled through a chute or using clamshell at a designated location where it solidifies. The technique's execution is similar to that of concrete placement. Soil and sand for ground improvement may be excavated in situ or transported from other sites. This means that not only in situ ground but also dredged material (from navigation channels) and construction waste soil can be effectively used. Sandy soils are used for ground improvement because clayey soils are difficult to handle. Moreover, the mechanical properties of solidified soil vary considerably according to the type of clay. If clayey soils are to be solidified and utilised, pneumatic flow or lightweight mixing (described in a later section) is employed instead of the pre-mixing method. The principle of soft soil improvement between shallow and deep mixing methods is the same. Solidification includes processes, such as the hydration reaction of solidifying materials, formation of hydrates, ion exchange reaction with clay minerals, and pozzolanic reaction. Details are presented in the section on shallow mixing method. However, note that the soils used in the pre-mixing method are sandy soils, which have a certain degree of stiffness and strength at the unimproved stage compared with clayey soils. For this reason, in implementing the pre-mixing method, only a small amount of solidifier is added to provide cohesion rather than incorporating a considerable amount to harden the soil completely. Hence, strength can be adjusted by varying the amount of solidifier added.

In coastal areas, pre-mixed soil is used as backfill behind quays and seawalls, filling material for steel plate cells, and replacement material after floor excavation. Soil solidification prevents liquefaction during earthquakes and reduces the earth pressure acting on quays and seawalls. This method was developed in Japan in the 1990s and initially used for strengthening the foundation ground of the tunnel section of the Trans-Tokyo Bay Expressway (Tokyo Bay Aqua-Line). The method also attracted particular attention during the restoration of quays and seawalls damaged by earthquakes. In 1995, a massive earthquake struck the southern part of Hyogo Prefecture (near Kobe City, Japan), severely damaging the Kobe Port facilities. Many of the quays were pushed out to sea, and the ground behind them was washed out. Eventually, the quays were restored, and the area behind them was reclaimed. Instead of filling the sections with soil (as they were), they were filled with cement-added soil prepared by the pre-mixing method. This method prevents ground liquefaction behind the quays and improves earthquake resistance by reducing the earth pressure on the quay wall. Further information on the pre-mixing method is found in Ref. [1].

4.1.2 Investigation

The required investigations for the pre-mixing method design include laboratory mixing tests to determine the type and amount of solidifier to be added to attain the required strength as well as to determine the properties of the soil to be used as the base material. The soil properties include physical characteristics, such as particle density, water content, particle size, and maximum and minimum densities. Density influences the weight of the solidified soil, whereas water and fine fraction contents influence the strength development after solidification. In addition to the base soil density, the density of the soil formed by the previous pre-mixing method must be used as a reference for soil density after solidification. Soils with similar grain sizes are presumed to have similar densities. If suitable data for estimating density after solidification cannot be obtained, field experiments can be conducted to derive them. Both density and strength are parameters required in design.

Laboratory mixing tests for the pre-mixing method differ to a certain extent from those for the shallow and deep mixing methods. For example, in the deep mixing method, the amount of solidifier added is considerable, and the strength of the cement-treated soil is high. Hence, laboratory mixing tests are fundamentally unconfined compression tests. In the pre-mixing method, the amount of solidifier and the strength of the cement-treated soil are moderate. Thus, in addition to the unconfined compression test, a tri-axial compression test is implemented. The pre-mixing ground improvement method is also expected to be an effective ϕ material. Furthermore, both cohesion (c) and internal friction angle (ϕ) must be determined. However, if the purpose of ground improvement is simply to prevent liquefaction, an unconfined compression test is sufficient. This is because liquefaction does not occur when the unconfined compressive strength is high. The liquefaction strength can also be determined based on the unconfined compressive strength. The relationship between the shear strength of the cement-treated soil and the type and amount of solidifier added must be determined by laboratory mixing tests.

4.1.3 Design

The design of the pre-mixing method differs from that of cement treatment methods including the deep mixing method. Ground strength is deemed to have increased to a limited extent, and the improved ground is treated as general soil similar to that in compaction methods. The external and internal stabilities are considered such that the solidified body does not move as a single unit or break because of the considerable strength attained by the improved ground via cement treatment including the deep mixing method. In contrast, the strength of the cement-treated soil in the pre-mixing method is moderate. Moreover, the design method differs from that of the deep mixing method because the improved ground is considered to have general soil characteristics.

The extent of improvement and strength of the cement-treated soil must be considered in design to ensure the stability of the structure. For example, if the pre-mixing method is used to reduce the earth pressure on the quay wall, the extent of improvement and the strength of the cement-treated soil must be determined for the earth pressure to satisfy the wall's required stability. Specifically, the stability is verified using an earth pressure equation that accounts for both cohesive force (c) and internal friction angle (ϕ), considering that c increases with the addition of the solidifying material. When the addition rate of the solidifier is 10% or less, the internal friction angle of the cement-treated soil is equal to or slightly exceeds that of the untreated soil. A safe value for the internal friction angle of the cement-treated soil is that of the untreated soil. Alternatively, a smaller internal friction angle, that is, 5–10°, obtained by the triaxial compression test may be used, or the internal friction angle may be estimated from the N value of the untreated soil. In addition to the stability of the quay wall against active earth pressure, circular slip analyses are conducted to check the stability of the entire wall structure. Such stability verification determines the strength required for the cement-treated soil (design strength).

The types and amounts of solidifier and anti-separation agent are specified based on the results of the laboratory mixing tests and design strength described above. The procedure is as follows. First, solidifiers, such as ordinary Portland cement and blast furnace cement, are selected. Next, the laboratory strength that demonstrates the design strength is determined. The design strength is obtained by multiplying the laboratory strength by a certain factor. This is because the strength exhibited in the field is lower than that obtained in the laboratory mixing test under satisfactory mixing conditions. The pre-mixing method does not account for the effect of variations in strength; it only rectifies the difference between laboratory and in situ strengths. This factor, called adjustment factor α, is expressed in the following relationship:

$$q_{uck} = \overline{q_{uf}} = \overline{q_{ul}} / \alpha, \tag{4.1}$$

where q_{uck} is the characteristic value of the design strength, q_{uf} is the in situ unconfined compressive strength, q_{ul} is the laboratory unconfined compressive strength, and α is the adjustment factor (ratio of unconfined compressive strengths). According to previous experiences, laboratory strength is typically greater than in situ strength, and $\alpha = 1.7$–2.6 is frequently used. If the water depth exceeds 5 m and a clamshell system is employed, $\alpha = 2.0$ (for liquefaction prevention) or $\alpha = 2.5$ (for earth pressure reduction) can be used. For large projects involving soil that has not been previously improved using this method, performing verification tests prior to the implementation of the main project is advisable to determine the adjustment factor. Once the laboratory strength is determined, the amount of solidifier to be added to achieve the necessary strength can also be determined based on the results of the laboratory mixing tests.

If liquefaction prevention is the objective, determining the strength of the cement-treated soil is also necessary. A significant relationship between the liquefaction and unconfined compressive strengths of cement-treated soil has been reported; treated soil with an unconfined compressive strength of 100 kN/m² or more does not liquefy. Based on the foregoing, the type and amount of solidifier must be identified and set such that the resulting unconfined compressive strength is 100 kN/m² or higher. If the unconfined compressive strength of the solidified soil is less than 100 kN/m², cyclic triaxial compression tests must be conducted to confirm that the soil does not liquefy.

When the pre-mixing method is applied (e.g., behind a quay wall), the stability of the solidified body must also be considered, assuming that it slides as a whole. As mentioned above, fundamentally, the pre-mixing method does not introduce the necessity for verifying external and internal stabilities. However, the possibility of the entire solidified body sliding cannot be precluded if the strength demonstrated at the site is greater than that expected; accordingly, only the external stability against sliding is verified. Figure 4.1 shows external forces acting on the quay wall and solidified body. The horizontal seismic intensity, which differs according to country or region, is described in the section on the deep mixing method. For example, in Japan, the seismic waveform at a site where a structure is to be constructed is estimated by assuming the epicentre, obtaining the seismic motion of the engineering bedrock, and accounting for the site amplification characteristics. The high-frequency component of this waveform is attenuated to obtain the maximum acceleration on the ground surface by one-dimensional response analysis. This acceleration is subsequently corrected for the duration of the earthquake to obtain the horizontal seismic intensity. The horizontal seismic intensity obtained in this manner is applied to the equilibrium equation to ascertain that the improved structure does not slide in the horizontal direction. Because the solidified body is unlikely to overturn as a whole or sink into the foundation ground, overturning and bearing capacities are not considered in the pre-mixing method. The partial coefficient of friction at the base of the solidified body can be set to 1.0. For the friction force at the base of the solidified body, the coefficient of friction can be set to 0.6 if the foundation soil is sandy, or the cohesive force can be determined if the foundation soil is cohesive. If sliding occurs, the study

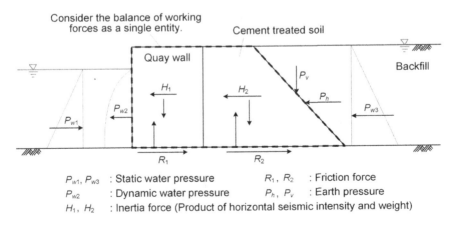

Figure 4.1 External forces acting on cement-treated body.

must be repeated until stability is achieved (e.g., by increasing the area of improvement).

4.1.4 Execution

The fundamental procedures for the pre-mixing method are as follows: (1) transport of landfill material; (2) mixing of soil with solidifier; (3) addition of anti-separation agents; (4) transport of cement-treated soil before solidification; and (5) placement of treated soil and reclamation. Several methods are available for mixing the solidifying material as well as placing and reclaiming the cement-treated soil. The appropriate method must be selected according to site conditions (e.g., type of base material, scale of execution, execution period, and economic efficiency). Actual mixing includes conveyor belt, self-propelled soil stabiliser, mechanical mixer, and rotary crusher–mixer systems. Figure 4.2 shows the execution system diagram for the rotary crusher–mixer method.

Execution management involves four items: quality control during execution, water quality control, water temperature control, and quality confirmation after execution. Quality control during execution includes material control (i.e., control of particle size distribution of soil, particle density, water content, and inspection results of solidifiers and anti-separation agents at the time of shipment), mixing control (i.e., control of supply amount of soil, solidifiers, and anti-separation agents as well as mixing status), and reclamation control (i.e., control of sediment shape when cement-treated soil is placed in water). Water quality control during execution focuses on pH and turbidity at representative points near the source, inside and outside the anti-pollution membrane, and inside and outside the execution area. The strength development of the cement-treated soil is considerably affected by seawater temperature; thus, water temperature must be monitored. In particular, the lower the temperature, the greater the mixture inadequacy, and the longer the delay of the initial onset of strength.

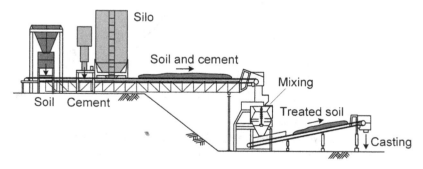

Figure 4.2 Execution system of rotary crusher–mixer.

For post-execution quality confirmation, check borings are implemented to determine the N value, density, and unconfined compressive strength through a standard penetration test (SPT). Seismic exploration, consolidation drainage triaxial compression tests, and calcium analysis are also implemented as necessary. The frequency of post-execution investigations is based on one point per 2,500 m^2 of ground improvement area at three depths (top, middle, and bottom) per point.

4.2 PRE-MIXING METHOD WITH SECONDARY MATERIALS

4.2.1 Overview

The pre-mixing method is introduced in the previous section. In this method, soil is improved by feeding cement or other solidifiers into the construction site. In addition to cement-treated soil, various by-products may be used as landfill materials. In coastal areas, these materials include construction by-products (e.g., concrete shells, asphalt, and construction waste soil), steel slag from steel mills, and incinerated ash from thermal power stations. Construction by-products and incinerated ash do not contain dangerous or environmentally hazardous substances; hence, they have been directly used as landfill material. Construction waste soil has also been used in the pre-mixing method. In terms of ground improvement, the use of steel slag has been mentioned. The methods for improving the ground with steel slag are introduced in this section.

Steel slag has been used as raw material for cement, aggregate for concrete, roadbed material for roads, and, more recently, as ground improvement material. In addition to its self-hardening feature, slag has the capacity to solidify the ground. Although steel slag cannot be expected to develop strength in the short term compared with cement, it solidifies starting from the medium term to long term. Hence, its use is effective for structures in which short-term strength is not required. Steel slag is broadly classified into blast furnace slag (produced when iron ore is melted and reduced in a blast furnace) and steelmaking slag (produced at the steelmaking stage when steel is refined). Steelmaking slag is further classified into converter and electric arc furnace slags. The blast furnace slag has been used as cement, concrete aggregate, and roadbed material. The steelmaking slag is hard and has been used as backfill material; however, its use has been limited because it contains unreacted CaO and expands when reacted with water.

In recent years, the solidifying properties of converter slag, which is classified as steelmaking slag, have been utilised to solidify soil. The calcium component in the converter slag reacts with the silica and alumina in the soil, resulting in soil hydration and solidification. The foregoing method for soil solidification was developed in Japan in the 2000s [2]. The soil stabilised with the converter slag as the main ingredient of the amendment is called calcia-stabilised soil (calcia means CaO (calcium oxide)). In particular, soft soils with high clay and silt contents (the fine fraction content exceeds 20%) derived by dredging navigation channels can be used as the main material to modify their physical and chemical properties. Calcia-stabilised soil is used not only to increase strength but also to control seawater turbidity and purify the seabed surface. It is applied to reclamation sites and used to fill navigation channels (fill material for areas adjacent to navigation channels), partition embankments, and shallow and tidal flat bases. In addition to the use of mixers and backhoes, pneumatic flow mixing and drop-mixing are employed to mix soil and converter slag.

Blast furnace slag alone (without soil mixing) is occasionally used as a geomaterial in coastal areas (note that this subject is outside the scope of soft soil improvement). Among blast furnace slags, granulated blast furnace slag is used for backfilling the quay walls and seawalls as well as lining soft ground. Granulated blast furnace slag is not only granular but also hydraulic and hardens with time. It has similar properties to slag-solidified soil. The quality of granulated blast furnace slag is stable when it is discharged from the same steelwork. The quality of improved soil is influenced by the physical properties of the original soil and extremely variable. Nevertheless, granulated blast furnace slag is more stable because it does not mix with soil. Because slag does not improve soil, it is not the subject of this book; however, its use as backfill is one of the most effective utilisation of this material in coastal areas.

4.2.2 Investigation

The investigations required for designing calcia-stabilised soil include the clear comprehension of the properties of the base soil as well as the properties of the converter slag (the main improvement material). Laboratory mixing tests are also conducted to determine the type and amount of solidifier to be added to derive the required strength. The soil properties include physical characteristics (such as particle density, water content, particle size, and maximum and minimum densities) and chemical properties (such as organic and hazardous substance contents). Density influences the weight of the improved soil, whereas water, fine fraction, and organic matter contents influence the strength development of the improved soil. The properties of slag include density, CaO content, and grain size. The CaO content and grain size of the slag affect the strength development of the improved soil. The density of the improved soil can be estimated not only by the density of the base soil and slag but also by the density of improved soils in previous works. The densities of soils with similar grain sizes are presumed to be similar. Density and strength are the parameters used in design.

As for laboratory mixing tests, unconfined compression tests are implemented in the same manner as those in the cement treatment method. Flow tests are also performed on the improved soil before solidification to check the workability and resistance to waves and currents. For strength, a low mixing ratio of slag shows a constraining pressure effect on strength, whereas a high mixing ratio reduces this effect. Typically, the mixing ratio of the slag is relatively high (10–40%) and is treated as c material without considering the confining pressure effect. The strength can be assessed by the unconfined compressive strength. In this method, the mixing ratio is fundamentally evaluated by volume rather than weight, and the ratio is calculated using the actual volume of the slag (excluding voids). As the slag contains relatively large grains, unconfined compression tests are performed using specimens as large as 100 mm (diameter) × 200 mm (height). In the past, a 30% slag mixing ratio has been observed to result in an unconfined compressive strength (28-day strength) of 50–400 kN/m². The mixing ratio of the slag in the mixing test must be determined with reference to the foregoing strength. The unconfined compressive strength at 7, 14, and 28 days of curing is investigated, considering three water content levels in the base soil and three mixing rate levels of the slag.

Flowability is examined by flow tests. Soil is placed in a cylinder with an open end before it solidifies. Then, the cylinder is pulled out, and the spread diameter of the soil is observed: The greater the fluidity of the soil, the wider the spread diameter. In Japanese flow tests, a cylinder with a diameter of 80 mm and height of 80 mm is used; the flow value must at least be 90 mm for this method. Mixers, backhoes, and drop mixing are used as execution methods when the flowability is low, and the pneumatic flow mixing method is employed when the flowability is high. Flow values are also used to check the degree of resistance to waves and currents.

4.2.3 Design

In the design of structures to be constructed on calcia-stabilised soil, ground strength is not extremely high such that the improved soil can be treated as relatively hard clay. For example, the stability of the ground including the improved soil can be evaluated using the bearing capacity equations for clay and conducting circular slip analyses to check slope stability. When calcia-stabilised soil is used to improve the environment of the seabed surface, checking whether the improved soil has not been washed away by water before it solidifies is necessary. For this verification, the results of flow or vane tests on unsolidified improved soil are used. The flow value must be less than and the vane shear strength must be greater than the corresponding permissible values, respectively.

The mixing ratio of the slag is determined based on the results of the laboratory mixing test to ensure that the strength required for the design is attained. The design strength is obtained by multiplying the laboratory strength by coefficients because the strength in the field is lower than that obtained in the laboratory under satisfactory mixing conditions. The coefficients consider the percent defective (the percentage of in situ strength below the design strength), variability (expressed as a coefficient of variation), and ratio of in situ strength to laboratory strength. The following relationships are established between the design strength and strength obtained by the laboratory mixing test:

$$q_{uck} = (1 - \alpha v)\overline{q_{uf}} \tag{4.2}$$

$$\overline{q_{uf}} = \beta\overline{q_{ul}}, \tag{4.3}$$

where q_{uck} is the characteristic value of the design strength, q_{uf} is the in situ unconfined compressive strength, q_{ul} is the laboratory unconfined compressive strength, α is the coefficient determined from the percent defective, v is the coefficient of variation, and β is the in situ–laboratory strength ratio. Coefficients α, β, and v must be determined based on the test results as they depend on the degree of mixing and stirring at the site and the execution condition. However, if the test execution is difficult to implement, the coefficients used in the pneumatic flow mixing method are typically used. This means that a percent defective value of 25% (with a normal deviation of 0.67), coefficient of variation of 0.35, and ratio of in situ strength to laboratory strength of 0.5 are considered. Actual execution results typically exhibit smaller coefficients of variation and larger in situ–laboratory strength ratios than the foregoing. Hence, a more favourable design may be achieved using the coefficients obtained by conducting test execution. Once the laboratory strength is determined, the amount of slag that must be added to achieve the necessary strength can be determined from the results of the laboratory mixing test.

The unit volumetric weight for design can be estimated by multiplying the surface dry density of the slag by the mixing ratio versus the density of the base soil. The improved soil can be treated as a non-liquefiable material. In addition, the method uses converter slag, which is virtually a non-expandable material because it is mixed with highly compressible soil. Environmental safety must also be checked. Specifically, the elution of hazardous substances and the effects of pH must be verified in accordance with national and regional standards.

4.2.4 Execution

The execution procedure for calcia-stabilised soil consists of transporting the materials, mixing the soil and slag, transporting the unsolidified improved soil, and placing the improved soil. Various mixing and feeding methods are available. They are selected according to the size, geographical conditions, soil conditions, and acceptance cycle of the soil. As summarised below and shown in Figure 4.3, four main types of mixing methods may be implemented.

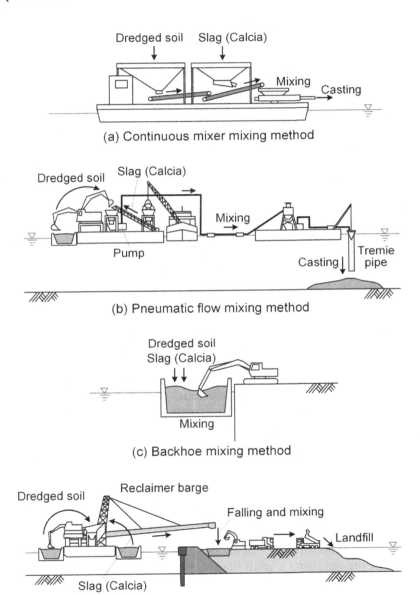

Figure 4.3 Mixing methods of calcia-stabilised soil.

1. Continuous mixer mixing method (execution capacity: 50–100 m³/h/ unit)

   This method involves mixing soil and solidifier using a continuous mixer on a work barge. It is suitable for small-scale execution; nevertheless, the execution capacity can be increased using several mixers. This method enables excellent mixing.

2. Pneumatic flow mixing method (execution capacity: 300–600 m³/h)

   This method uses slag instead of cement as solidifier (the use of cement in the pneumatic flow mixing method is described in the next section). The soil and improvement material are mixed by the turbulent effect of the plug flow generated in the pneumatic feed pipe. This method is suitable for large-scale execution. It also enables excellent mixing.

3. Backhoe mixing method (execution capacity: 50–200 m³/h)

   This method involves putting the solidifier and soil in an earthmoving vessel and then mixing them using a backhoe. Efficient mixing can be achieved by attaching an agitator to the backhoe. This method is suitable for small-scale execution.

4. Falling mixing method (execution capacity: approximately 600 m³/h)

   This method mixes the soil and solidifier by impact during transfer via the conveyor belt of the reclaimer vessel or when they fall from the spreader. The mixing accuracy is adjusted by modifying the drop height and point. The method is suitable for large-scale execution.

The three main feeding methods (Figure 4.4) applied are as follows.

1. Direct feeding method (bottom-opening barge feeding)

   This method involves feeding improved soil directly into the sea from bottom-opening barges. Although the technique is economical, controlling the shape of the material is difficult after soil placement. Moreover, the strength of the material considerably varies due to seawater mixing when the soil is dropped into the sea. This method is suitable for sites with shallow water depths.

2. Tremie pumping method

   This method involves pumping improved soil into the seabed via tremie pipes. The material is less prone to seawater entrainment and to the influence of sea turbidity; hence, the strength variation is low. However, to pump the improved material, a certain degree of soil fluidity is required before solidification.

3. Grab feeding method

   This method involves placing the improved soil loaded onto a barge into the sea using a grab bucket. By lowering the grab bucket near the seabed, turbidity and other problems can be controlled, and soil strength variations are reduced. However, the technique's execution efficiency is low; hence, the method is only suitable for small sites.

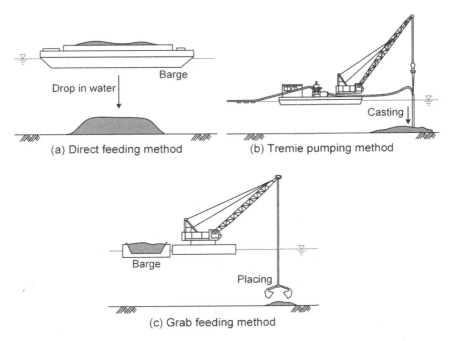

Figure 4.4 Feeding methods of calcia-stabilised soil.

For effective execution management, material (soil and solidifier), mixing (density), placement (position, quantity, etc.), quality (strength), and environmental effects (turbidity, pH, etc.) must be controlled. In the case of the backhoe mixing method, the relationship between the mixing time and wet density of improved soil must be determined, the degree of variation must be checked, and the mixing time must be set to minimise variation. For quality control after execution, the strength of the ground in general must be checked by soundings, such as cone penetration test (CPT) and SPT. In some cases, sampling and unconfined compression tests may be implemented. These tests verify that the number of points below the required strength is less than the percent defective. In addition, the workmanship of improved soil placement in terms of ground height, length, width, and slope is controlled by depth surveying.

4.3 PNEUMATIC FLOW MIXING METHOD

4.3.1 Overview

In coastal areas with ports and harbours, navigation channels are typically dredged, and sediment is temporarily stored in sediment disposal sites.

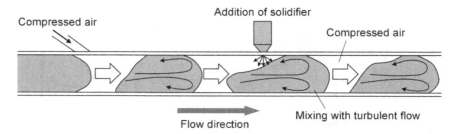

Figure 4.5 Agitation in pipe of pneumatic flow mixing method.

Because the capacity of these sites is limited, large quantities of sediment may be moved to another location for effective use. However, dredged sediment often has a high clay/silt content; hence, it may be difficult to utilise in its existing state. In such cases, an effective ground improvement method is pneumatic flow mixing. In this method, sediment is pneumatically pumped through pipes. During pumping, cement or other solidifiers are added and mixed, as shown in Figure 4.5. The soil and solidifier are stirred using the turbulent effect of the plug flow generated in the pumping pipe. In other techniques, the soil and solidifier are blended at the plant using a mixer, rendering the solidification process less efficient. However, in the pneumatic flow mixing method, the soil is continuously transported, and the solidifier is automatically mixed with the soil during the process; consequently, the solidification process is facilitated. The principle of soil solidification in pneumatic flow mixing is the same as that in the shallow and deep mixing methods. Solidification proceeds through the hydration reaction of the solidifying material, formation of hydrates, ion exchange reaction with clay minerals, and pozzolanic reaction. This method is applied to cohesive soil with less than 50% sand and gravel content. Pneumatic pumping is most suitable for clayey soils with a sand and gravel content of 30% and natural water content ratio of 1.3–1.5 times the liquid limit (w_L) of the soil.

Improved soils may be used as landfill, backfill (behind seawalls), and embankment (for partitioned seawalls) materials. They are also placed immediately behind quays and seawalls to reduce earth pressure and prevent liquefaction. This method was demonstrated at the Port of Nagoya, Japan, in 1998 and subsequently used extensively in the reclamation of the airport island at Central Japan International Airport and Tokyo International Airport. In recent years, the method has also been adopted for coastal works in North America and Southeast Asia and is currently being used worldwide. The details of the pneumatic flow method are well summarised in Ref. [3].

4.3.2 Investigation

The design of the pneumatic flow mixing method requires investigations, including laboratory mixing tests, for determining the type and amount of solidifier to be added. This enables the attainment of the required strength and fluidity as well as determines the properties of the soil as base material. Hexavalent chromium leaching tests are also implemented to investigate environmental effects. The physical properties of the soil, such as particle density, water content ratio, particle size, and liquid and plastic limits, are determined. Density influences the weight of the cement-treated soil, whereas information on water content, sand and gravel contents, and liquid limit is important when considering the applicability of the method. Because the weight of cement-treated soil is a parameter used in design, the density of the ground formed by previous pneumatic flow mixing methods must also be considered in addition to the density of the base soil.

In the laboratory mixing tests for the pneumatic flow mixing method, other tests must be implemented in addition to the strength tests for the shallow and deep mixing methods. In pneumatic flow mixing, ascertaining that the soil can be pneumatically pumped into the pipe before solidification (that is, the soil has sufficient fluidity for pumping) is necessary. Strength is assessed by unconfined compression tests similar to other cement treatment methods. The water content of the base material suitable for the pneumatic flow mixing method is approximately 1.3–1.5 times the liquid limit. With this water content, the reference value of cement to be added per unit volume of soil is approximately 50–100 kg. The addition of cement must at least be in three levels, and the unconfined compressive strengths at 7 and 28 days of curing must be examined. Flowability is investigated by the flow test wherein soil is placed in a cylinder with open ends before solidification. Then, the cylinder is pulled out, and the spread diameter of the soil is observed. If long-distance pumping is required and a cylinder with a diameter and height of 80 mm is used, a flow value of at least 90–100 mm with an upper limit of 200–250 mm must be attained. The flow value can be adjusted according to the water content ratio; in actual execution, this can be achieved by modifying the amount of water added. During the casting process, the slope can be controlled according to the difference in flow value. If a large slope is desired, a small flow value must be used. If the frequency of the tip movement of the cylinder discharging the cement-treated soil from the placing machine is to be reduced, a high flow value must be used to make the slope small. However, if the water content is overly high, the amount of solidifier required must be increased; otherwise, material separation is certain. Accordingly, an upper limit is specified.

4.3.3 Design

This section describes the design in the use of soil improved by pneumatic flow mixing. The approximate strength achieved by the soil improved using this method is 100–500 kN/m², which is lower than that achieved using deep mixing. For a hard clay layer, fundamentally the same design methods can be employed. For example, the stability of the ground (including the cement-treated soil) can be evaluated using the bearing capacity formula for clay soils and conducting circular slip analyses for slope stability. However, when the pneumatic flow mixing method is employed behind a quay wall, a relatively special method is used to calculate the earth pressure acting on the wall due to the solidified soil. This is because the existing earth pressure equation assumes that the ground behind the wall is semi-infinite, the layer thickness is constant, and the layers in the soil are parallel. When cement-treated soil is used behind a wall, its length from the wall is finite, and an earth pressure calculation formula must be used to account for this length. In this case, the split method is applied to calculate the active earth pressure. For the earth pressure acting on the wall due to the treated soil, a slip surface (usually a straight slip surface) is assumed behind the wall. The soil mass sandwiched between the slip surface and wall is divided horizontally. The active earth pressure is calculated by establishing the equations for the weight, buoyancy, shear, and seismic horizontal force acting on each of the fragments. In the case of ground with cement-treated soil above and below it, four modes of slip failure are assumed, as shown in Figure 4.6. The earth pressure in each mode is calculated up to the upper limit of the active earth pressure. In calculating the active earth pressure, the frictional force between

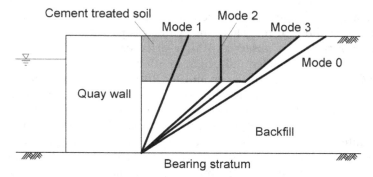

Mode 0: Sliding behind cement treated soil
Mode 1: Linear slip through cement treated soil
Mode 2: Sliding with crack
Mode 3: Sliding along cement treated soil

Figure 4.6 Four modes for calculating active earth pressure.

the solidified soil/cohesive soil and wall surface is ignored. In contrast, the frictional force between the sandy soil and wall surface is considered. The wall friction angle is set to 15°.

The type and amount of solidifier to be added are determined according to the results of the laboratory mixing tests. This ensures that the strengths required by the design and fluidity of the cement-treated soil during pumping are attained. The procedure is as follows. First, solidifiers (cements, such as ordinary Portland cement and blast furnace cement) are selected. Next, the laboratory soil strength that can demonstrate the design strength is determined. The design strength is obtained by multiplying the laboratory strength by a factor. This is because the strength in the field is lower than that obtained in the laboratory by mixing test conducted under excellent mixing conditions. The coefficients account for the percent defective of solidified soil (i.e., the percentage of in situ strength less than the design strength), variability (considered as a coefficient of variation), and ratio of in situ strength to laboratory strength. Fundamentally, the percent defective is considered as 25% (with a normal deviation of 0.67), the coefficient of variation is 0.35, and the ratio of in situ strength to laboratory strength is 0.7 (in air) or 0.5 (in water). Hence, the design and laboratory strengths are related by the following:

$$q_{uck} = (1 - 0.67 \times 0.35)\overline{q_{uf}} \tag{4.4}$$

$$\overline{q_{uf}} = 0.7 \text{ or } 0.5 \times \overline{q_{ul}}, \tag{4.5}$$

where q_{uck} is the characteristic value of the design strength, q_{uf} and q_{ul} are the in situ and laboratory unconfined compressive strengths, respectively. Ideally, the coefficient of variation and in situ–laboratory strength ratio must be obtained by prior test execution. However, if this is not possible, the values specified above can be used. Once the laboratory strength is determined, the amount of solidifier to be added to achieve such strength can be determined from the results of the laboratory mixing test. Then, checking whether the soil can maintain its fluidity before solidification is necessary. As mentioned, the flow value in a cylinder with a diameter and height of 80 mm must at least be 90–100 mm and not exceed 200–250 mm.

4.3.4 Execution

The execution process for pneumatic flow mixing consists of mud thawing, water addition, pneumatic pumping, solidifier addition, and feeding. Two types of execution methods are possible: offshore execution (implemented from sea using special work vessels) and onshore execution (implemented by installing soil pumping and solidifier supply equipment on land). The offshore execution system is shown in Figure 4.7. As shown in the figure,

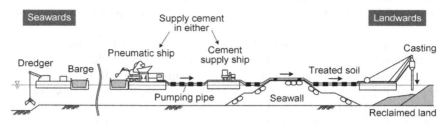

Figure 4.7 Execution system of pneumatic flow mixing method.

the solidifier can be added either by the pumping system or line addition system. Note that maintaining a sufficient pipe length from addition to discharge to achieve adequate mixing by plug flow is important. The main feeding techniques are slope flow-down, pumping, tremie barge, and direct feeding methods. When the treated soil is to be placed in water, the slope flow-down method or pumping method must be adopted to minimise soil quality deterioration and the impact of the treated soil on the surrounding environment due to material separation and seawater entrainment.

The execution management of pneumatic flow mixing is implemented considering the control of material (dredged material and solidifier), pumping, placement, quality, and environmental effects. Ensuring that the quality required by the design and the amounts of solidifier and water to be added are achieved are particularly important. When the scale of execution is large, the continuous measurement of the water content of the dredged soil is possible using a gamma-ray densitometer such that the amount of solidifier to be added may be adjusted. When the execution scale is small, the water content of the dredged soil is measured for each earthmoving vessel and then the amount of solidifier to be added is determined. Because a close relationship exists between the strength of the solidified soil and water–cement ratio, daily feedback on the results of strength control with respect to the mix is advantageous.

For post-execution quality control, soundings, such as CPT or sampling of the cement-treated soil, are implemented to determine the soil's unconfined compressive strength. The examination of soil strength 28 days after casting is a common practice. In addition, after the treatment execution, the extent of casting (e.g., in terms of height, length, and width) and slope are measured to control the workmanship.

4.4 LIGHTWEIGHT MIXING METHOD

4.4.1 Overview

One effective method for dredged soil (excavated in coastal areas) and construction waste soil is lightweight mixing [4]. In this technique, the soil, water, lightweight material (foam), and solidifier are blended in a plant mixer or similar machine. Water is added to the base material such that the water content ratio exceeds the liquid limit. The formation of ground with lightweight cement-treated soil becomes possible by mixing foam; this is called the Super Geo-Material method. The unit volumetric weight can be modified by adjusting the amount of foam added, generally 8–13 kN/m³. However, if the unit volumetric weight is less than that of seawater, the material will float below the water level. Thus, the unit volumetric weight must exceed that of seawater when used below the water level. Expanded poly-styrol blocks and foamed mortar are widely used as lightweight embankments in land civil engineering structures. However, they are less commonly used in coastal areas because they are extremely light and float in water. They may be used if the overburden load is adjusted such that it exceeds the buoyant force. However, cases wherein structures have been damaged and cannot be repaired due to tidal levels and waves have been reported. In addition, many civil engineering structures in coastal areas are large in scale, and cost concerns must be considered. The lightweight mixing method, which enables the adjustment of unit volumetric weight to produce lightweight and high-strength soil, is widely used in Japan. Because of the advantages of using lightweight soils, mountain soils are occasionally utilised as base materials in addition to soft soils.

The principle of soil solidification in lightweight mixing is similar to those of the shallow and deep mixing methods. Solidification proceeds by hydration reaction and hydrate formation of the solidifying material, ion exchange reaction with clay minerals, and pozzolanic reaction. The lightweight mixing method can be applied to a wide range of soil types and can be used to improve clayey to sandy soils. However, when improving soils with a sand content of 50% or more, fine fraction, such as bentonite, must be added to increase fluidity. Figure 4.8 shows a magnified image of

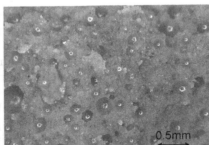

0.5mm

(a) Before solidification   (b) After solidification

Figure 4.8 Cement-treated soil produced by lightweight mixing method. (Courtesy of TOA Corporation.)

the cement-treated soil produced by the lightweight mixing method; the bubbles are not continuous (assuming a non-continuous bubble addition rate). Water does not easily penetrate the bubbles underwater. The hydraulic conductivity of cement-treated soil is extremely low and has been found to remain lightweight in the long term.

Improved soils are used as landfill and backfill materials behind seawalls and to counteract liquefaction. Because the soil improved by lightweight mixing is not heavy, the technique is applied to port and harbour structures to reduce the settlement of ground beneath the improved soil and the earth pressure on quays and seawalls. The first case in which this method was applied on site was in the quay restoration work at the Port of Kobe following the 1995 Hyogo-ken Nanbu earthquake in Japan; recall that the pre-mixing method was also used in this project. In this case, the quay had been damaged prior to the completion of backfilling works. Moreover, although the quay had been displaced due to the seismic force, the decision was to use the quay as it was. The plan was to design and construct a quay wall with improved earthquake resistance by introducing an embankment on the seaward side and lightweight cement-treated soil at the rear. The ageing of the treated soil is regularly monitored until this time, and although surface deterioration is observed, the function of the cement-treated soil as a whole is confirmed to have been maintained.

4.4.2 Investigation

The investigations required for designing the lightweight mixing method include laboratory mixing tests for determining the type and amount of solidifier to be added to achieve the required strength and fluidity as well as to determine the properties of the base soil. Hexavalent chromium leaching tests are also implemented to investigate environmental effects. The physical properties of soil, such as particle density, water content ratio, particle size, liquid limit, and plastic limit, are determined. Density influences the weight of cement-treated soil. Information on water content, sand and gravel content, and liquid limit is important when considering the applicability of the method.

Laboratory mixing tests are also implemented for the lightweight mixing method. In these tests, checking not only the soil strength developed but also whether the soil has sufficient fluidity such that it can be pumped is necessary. This is because the soil mixed with cement and lightweight materials is pumped through the pipe before feeding. Checking the unit volumetric weight is particularly important because part of the objective is to reduce weight. The viscosity of the material must also be checked to ensure that the treated soil does not separate in seawater when it is cast. As with other solidification methods, the strength of the solidified material is assessed by unconfined compression tests. Strength is tested by mixing soil, water, cement, and lightweight materials. If the base material is clayey soil, the water content must be approximately 2.5 times the liquid limit. If the base material is sandy soil, the water content must be approximately 120%. Cement-treated soil prepared with a mixing rate of 50–200 kg/m³ per soil volume is used as reference. The unconfined compressive strengths at 7 and 28 days of curing considering three levels of water content and solidifier addition are investigated. Flowability is investigated by flow test. The soil is placed in a cylinder (80 mm (diameter) × 80 mm (height)) with open ends before solidification and then the cylinder is pulled out to observe the spread diameter of the soil. Typically, the water content ratio and amounts of cement and foam are set such that the flow value is in the range 130–230 mm. The density must be considered because it increases by 0.05 g/cm³ as a result of the water absorption of the surface exposed to seawater, defoaming during execution, and shrinkage of the cement-treated soil.

Viscosity must be investigated by conducting an underwater separation resistance test. This test is set up independently by the developers of this method to ensure that the unit volumetric weight is not reduced or the water is not clouded by the separation of unsolidified treated soil in water (e.g., during soil casting). In the test, 3 ℓ of the unsolidified soil is placed in a cylinder with an inner diameter of 100 mm and a height of 440 mm and pushed out by a piston at a speed of 100–1000 mm/s. The soil is poured into a cylindrical vessel with an inner diameter of 200 mm containing 3 ℓ of artificial seawater. One minute after all the soil has been poured, the supernatant liquid in the receiving vessel is collected, and the amount of suspended solids (SSs) and pH are measured to confirm that they are less than 100 mg/ℓ and 10.5, respectively. The cement-treated soil is left to cure for seven days before it is collected as a specimen. Upon collection, its density, water content, and unconfined compressive strength are determined. Compared with the specimens prepared in air, the difference in density must be within ±0.05 g/cm³, the increase in water content must be less than 10 %, and the increase in unconfined compressive strength must exceed 50%.

4.4.3 Design

This section describes the design of soils improved by the lightweight mixing method. The approximate strength of the cement-treated soil achieved by this technique is 100–500 kN/m² (frequently 200 kN/m² in practice), and the unit volumetric weight is typically adjusted to 11.5–12.0 and 10.0 kN/m³ underwater and above the water surface respectively. Compared with the strength of soil improved using the deep mixing method, the strength of soil prepared by lightweight mixing is not as high; this is common in pneumatic flow mixing method described above. Fundamentally the same design methods are applied to hard clay layers. The stability of the ground (including the cement-treated soil) can be evaluated using the bearing capacity formula for cohesive soils and conducting circular slip analyses for slope stability. However, when the lightweight mixing method is implemented behind a quay wall, the earth pressure equation can be applied to the ground containing a finite length of cement-treated soil to calculate the earth pressure acting on the cement-treated soil up to the wall. As in the calculation of earth pressure using the pneumatic flow mixing method, the equation for the active earth pressure based on the division method is used. To determine the earth pressure from the solidified soil to the wall, a slip surface (usually a straight slip surface) is assumed behind the wall. The soil mass sandwiched between the slip surface and wall is divided horizontally. Then, the active earth pressure is calculated by establishing the equations for the weight, buoyancy, shear, and seismic horizontal force acting on each of the divided segments. In the case in which treated soil is above and untreated soil is underneath, four modes of slip failure are assumed, and the maximum active earth pressure in each mode is calculated, as shown in Figure 4.6. In calculating the active earth pressure, the frictional force between the solidified soil/cohesive soil and wall is ignored. In contrast, the frictional force between the sandy soil and wall is considered. The wall friction angle is set to 15°.

The water content ratio of the base material, the type and amount of solidifier added, and the amount of lightweight material (e.g., foam) added are based on the laboratory mixing test results to ensure that the design strength, unit volumetric weight, fluidity during the pumping of cement-treated soil, and required viscosity during placement are satisfied. The procedure is as follows. First, solidifiers are selected from cements (such as ordinary Portland cement and blast furnace cement). A foaming agent suitable for the soil material and cement is also selected. The main types of foaming agents are surfactant-based and animal protein-based. With regard to the amount of foam added, the volume reduction of the foam must be considered because shrinkage and defoaming occur due to pressure changes and friction in the pipe during mixing, pumping, and placing. Pressure-induced defoaming is calculated according to Boyle's law. Defoaming during execution is managed by increasing the amount of foam actually mixed at the site. Next, the laboratory strength of treated soil that can reflect the design strength is determined. The design strength is obtained by multiplying the laboratory strength by a factor because the soil strength in the field is lower than the strength given by the laboratory mixing test conducted under satisfactory mixing conditions. In the lightweight mixing method, the difference between the laboratory and in situ strengths is corrected by a single coefficient that accounts for variations in strength and other factors. This adjustment factor, called the adjustment factor, α, is expressed in the following equation:

$$q_{uck} = \overline{q_{ul}} / \alpha, \tag{4.6}$$

where q_{uck} is the characteristic value of the design strength, q_{ul} is the unconfined compressive strength in the laboratory, and α is the adjustment factor (ratio of unconfined compressive strengths). Based on previous knowledge, the laboratory strength is greater than the in situ strength, with $\alpha=$ 2.2 as the standard value. This value is the ratio when the percent defective (the percentage of in situ strength below the design strength) of the solidified soil is 15.9%, the coefficient of variation is 0.35, and the ratio of in situ strength to laboratory strength is 0.7. In the past, the coefficient of variation used is in the range 0.25–0.34, and the ratio of in situ strength to laboratory strength is in the range 0.59–0.88. The ratios are set with reference to these values. Once the in situ strength is known, the amount of solidifier to be added to achieve such strength can be determined from the results of the in situ mixing test. Considering the amount of solidifier added, ascertaining that the soil can maintain its fluidity before solidification is necessary. As mentioned, the flow value in a cylinder with a diameter and height of 80 mm must be approximately 130–230 mm. If the base material is sandy soil (with a sand content of more than 50%), fine fraction, such as bentonite, must be added as thickener to improve flow. The separation resistance in water must also be checked.

4.4.4 Execution

The lightweight mixing method involves water intake and pre-treatment, mud thawing, mud preparation, kneading, pouring, and curing. Continuous and batch-type mixing facilities are necessary; the appropriate facility must be selected according to the required production capacity. When large quantities of lightweight cement-treated soil are required, the necessary equipment can be installed on board a barge, or a special-purpose vessel may be used. During mud thawing and preparation, water is added to adjust the water content and remove gravel, stone, and other components larger than 5–10 mm. The soil is mixed with the solidifier and lightweight material (foam) produced on site by a foaming machine. The lightweight cement-treated soil produced is typically pumped to the site and placed with special tremie pipes. A typical execution system for a production capacity of 100 m³/h of treated soil is shown in Figure 4.9. In cases where no records must be referred to or no special conditions must be considered for execution (such as soil placement at considerable water depths or the use of special raw soil), conducting a trial test to confirm workability prior to the main execution is advisable.

The management of the execution of lightweight mixing method involves the control of material (dredged soil, solidifiers, lightweight materials,

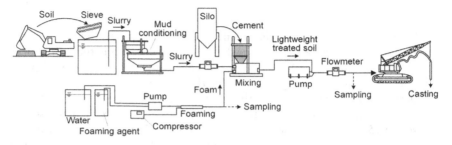

Figure 4.9 Execution system of lightweight mixing method.

water, and additives), pre-treatment, mud preparation, and mixing (density and flow value), placement (placement quantity and workmanship), quality, and environment (water quality). In terms of quality control during execution, in addition to strength, the control of the density and flow value of the cement-treated soil is particularly important. This is because the expected improvement is not only in terms of strength but also performance as a lightweight material. The flow value is also an important aspect in terms of workability. For quality control after execution, sampling is implemented to determine the unconfined compressive strength, density, and water content. To determine the unconfined compressive strength, the typical practice is to implement one sampling (at three locations: top, middle, and bottom) per 2500 m² of improved area. If necessary, in situ tests, such as CPT, may be performed.

4.5 FILTER PRESS OF SOFT CLAY

4.5.1 Overview

In coastal areas, dredging works are continuously conducted to maintain sufficient depths for navigation channels and anchorages. Dredged clay typically has a high water content exceeding the liquid limit due to water addition during dredging, rendering the dredged soil difficult to use. To provide the soil with satisfactory quality, the concept is to solidify it with cement or slag, as described in the previous sections. Consolidation is another possible approach that can be implemented. The soil is consolidated by the load applied in situ. After consolidation, the soil is dug and utilised. Although this technique is reliable, it is costly and time-consuming as well as requires soil to consolidate. Moreover, the load is not considerable such that high-strength soil is difficult to produce. Accordingly, a technique that can accelerate consolidation and dewatering—called filter press—has been devised. In this method, high consolidation pressure, that is, 1–4 MN/m₂, is applied to dewater dredged clay, resulting in soil volume reduction (approximately 20–50%) and high strength (cone index, q_c, is 400–600 kN/m^2 or more).

The dehydration of materials using filter press has been practiced since the BCE (before the Common Era) period using wooden filters to extract wine, oil, and other substances. A major turning point was during the Industrial Revolution in the 1800s when mechanical filter presses with filters and frames made of iron were developed in Great Britain. Mechanical dewatering machines were heavily used to dewater mud in coal mines. The further development of this method was triggered by the invention of the single filter plate and fully automatic filter press by Japanese Kurita Machine Works in 1959. The single filter plate is a concave plate that eliminates the necessity for a frame to produce a dewatered cake (a mass of soil after dewatering). The mud was filtered by placing it in a filter cloth, which could be vibrated after dewatering to remove the dewatered cake from the filter by its own weight without having to rely on human labour. Since then, some companies have endeavoured to develop new technologies. Currently, the filter press is mainly used in coastal areas to reduce the volume of dredged clay and facilitate handling. However, the progress in effectively utilising dewatered soil is not significant; it has only been used as reclamation material or trial embankment material. This is presumed to be because the soil properties of dewatered soil remain unknown thus far, and sufficient data on soil constants required for design have not been derived. If the properties of dewatered soil are clarified, it may be used for other applications in the future.

4.5.2 Investigation and Design

To design a filter press, investigating the properties of the soil to be used as base material is essential. Physical properties, such as soil particle density, water content ratio, grain size, and liquid and plastic limits must be determined. The density and water content ratio are necessary to control the water content ratio when water is added. The particle size indicates whether the soil contains sand or gravel, which must be removed. The liquid and plastic limits must be determined because soils with a liquid limit of 70–100% and plasticity index of 40–70 are regarded suitable for filter pressing.

In designing civil engineering structures to be built on mechanically dewatered soil, the improved soil can be treated as relatively hard clay soil, similar to soils improved using other ground improvement methods. For example, the stability of the ground including dewatered soil can be evaluated using the bearing capacity equation for clay soils and conducting circular slip analyses for slope stability. The investigation of unit volumetric weight, strength, consolidation properties, and hydraulic conductivity of the dewatered soil may be necessary. In the ground improvement methods described thus far, the soil's physical properties may be adjusted to some extent during the ground improvement process. For example, in the compaction method, strength can be modified by varying the degree of compaction. In the cement treatment method, the strength can be adjusted by varying the amount of cement added. To ensure the stability of the structure, the required physical properties of the soil can be determined, and to derive these properties, ground improvement specifications can be identified. In contrast to these ground improvement methods, the physical properties of the soil in the filter press are difficult to adjust. Furthermore, the properties of dewatered soil have not been fully elucidated. Therefore, the implementation of test execution to understand the characteristics of the prepared dehydrated soil or to design the soil with provisional soil constants (which should be revised if the assumed soil constants differ from those in the execution) is necessary.

The unit volumetric weight of dewatered soil varies with compaction. The dewatered soil produced by the filter press consists of slabs that are piled up forming a subgrade, as shown in Figure 4.10. Therefore, determining not only the unit volumetric weight of the clods but also the unit volumetric weight of the ground on which the clods are piled up is important. The compaction above the water level using execution equipment is possible. The wet unit volumetric weight, γ_t, of the compacted soil is approximately 16.5 kN/m^3; however, because it cannot be compacted underwater, γ_t has been proven to be approximately 14.5 kN/m^3. Evidently, the unit volumetric weight varies with the degree of compaction. As regards to strength, c_{cd} and ϕ_{cd} have been found equal to 5 kN/m^2 and 30°, respectively. The common practice is to consider the shear strength of compacted soil as that of ϕ material. This is because the pore space of piled-up soil masses forming the ground is large, and the shear resistance angle varies according to the

Figure 4.10 Soil treated by filter press.

confining pressure. Note, however, that the shear properties may vary as the degree of compaction on the land above the water surface is increased. Tests have indicated that the consolidation coefficient and hydraulic conductivity are approximately 10,000 cm²/d and 0.001 cm/s, respectively.

4.5.3 Execution

The mechanical dewatering process by the filter press involves placing a slurry of cohesive soil in a permeable bag (called a 'filter cloth') in a filter chamber. The slurry is continuously injected at high pressure for dewatering and compaction. The filter press consists of the following: pre-treatment (to remove sand and gravel), mud storage, reaction (mixing tank for auxiliary materials), high-pressure dewatering treatment, and water treatment. Some filter press facilities are plate-shaped machines loaded to dewater slurry by applying pressure. However, slurry dewatering by applying pressure is more common in ground improvement methods. For example, the thickness of dewatered soil in a filter chamber (120 cm × 120 cm in width and height, respectively) is 23.5 mm. The device has 122 filter chambers with an overall volume of 3.2 m³.

Figure 4.11 shows the filter chamber structure and dewatering process. First, sand, gravel, shells, and other materials larger than 0.075 mm that interfere with the dewatering process are removed by a trommel or other means. These materials in the soil clog the shaft of the slurry pump; they interfere with dewatering or cause the wear of the filter cloth and other parts of the pump. The water content ratio of the soil is then increased (e.g., 300–500%) such that slurry is formed. Moreover, a small amount of poly-aluminium chloride or slaked lime is added as dewatering aid. The slurry is

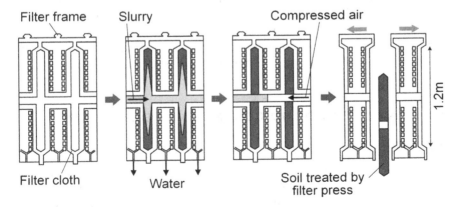

Figure 4.11 System and execution process of filter press.

fed into a filter chamber equipped with a filter cloth using a pump with a high flow rate. The pressure is approximately 0.7 MN/m$_2$. A high-pressure hammering pump is then used to increase the pressure to 4 MN/m$_2$ for consolidation dehydration lasting for approximately 1 h from the start of dewatering. After dewatering, the resulting water content ratio is 50–150%.

The execution management of the filter press must include identifying the soil characteristics, controlling the sediment concentration, determining the amount of dewatering aids to be added, identifying the characteristics of dewatered soil, and controlling the filter water quality. Moreover, the sand and gravel in the sediment must be removed as described above. The sediment concentration and amount of auxiliary material added for dewatering are adjusted using the automatic operation control system of the equipment. However, in some cases, the slurry concentration is regulated by cylinder sedimentation tests using samples. The amount of auxiliary material added is controlled by measuring the specific gravity (mud balance) in the sludge storage tank. The quality control immediately performed after execution involves measuring the water content of the dewatered soil and conducting in situ surveys, such as CPT and SPT. In some cases, sampling is conducted, and unconfined compression, triaxial compression, simple shear, and consolidation tests are performed in the laboratory as necessary.

Long-term quality control after execution is also important. This is because dewatered soil that has not been solidified by cement or other means becomes fine-grained after repeated dry and wet cycles. Thus far, quality control methods have not been established. At the Shin Moji offshore sediment disposal site in Japan (near Kitakyushu Airport), for example, the strength of temporary embankments constructed from dewatered soil had been assessed. The assessment was conducted using CPTs incorporating radioisotopes, surface wave surveys, and laboratory mechanical tests after block sampling. The strength of the temporary embankment made of dewatered soil is evaluated over time. In this case, the strength of the near-surface layer decreased ten years after the temporary seawall was constructed. The stability of the entire seawall must also be checked over time in the future. The technique of mixing a small amount of cement before using the filter press to achieve long-term soil stabilisation is also undergoing development.

4.6 USAGE OF GROUND CEMENTED BY DEEP MIXING

4.6.1 Overview

In Chapter 3, the deep mixing method is introduced as a method for improving the soil properties of the original ground. However, the cement-treated soil produced by deep mixing can be more actively used as a civil engineering structure component. This is because the resulting strength of cement-treated soil is considerably high. For example, cement-treated soil is used in the construction of quay walls or as core material inside levees. The foregoing is introduced in this section. Because the general investigation, design, and execution of deep mixing have already been described in Chapter 3, this section presents an overview of the method. Moreover, it discusses the difference of the foregoing method from the deep mixing method applied to original ground.

4.6.2 Quay Wall Construction Using Deep Mixing Method

The typical structural quay walls in coastal ports and harbours include gravity (using caissons), sheet pile (using steel pipes), and pier (combining steel pipes and slabs) types. The structural type most applicable to the site has been selected, and all types have been widely adopted. Moreover, more economical methods for constructing new quays and modifying existing quays are required. One technique was proposed and adopted in Japan. The method involves solidifying the ground by deep mixing and then cement-treated soil is used as the main body of the quay (earth retaining structure behind the ground), as shown in Figure 4.12. When an ageing sheet pile quay wall is to be repaired or requires deeper embedment in conjunction with the repair, casting new sheet piles or reconstructing the quay wall as a

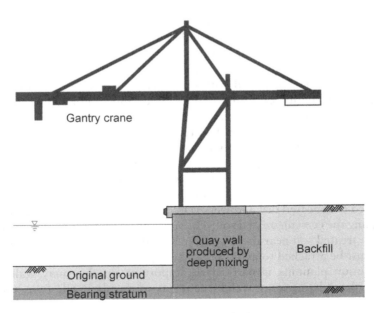

Figure 4.12 Quay walls produced by deep mixing method.

gravity or pier-type quay wall is necessary, requiring considerable budget. In this method, the existing sheet piles are left in place, and the ground behind them is solidified, enabling their use as earth retaining walls. This improvement approach is significantly less expensive than the conventional quay wall improvement method. In developing the method, the objective is to economically repair sheet pile quays; however, since the method's introduction, it has been widely adopted as a structural form for quays to sustain heavy loads. Specifically, these quays are required for the construction and maintenance of offshore wind power facilities. In countries where few earthquakes occur, foundation piles are typically driven into the ground to support heavy loads. In contrast, in countries and regions with frequent earthquakes (such as Japan), supporting the piles is difficult due to the plasticity of the foundation caused by horizontal seismic forces. Therefore, the idea of constructing the quay wall and supporting the heavy load using cement-treated soil has been adopted.

The investigation process for forming the quay using the deep mixing method is fundamentally the same as that for improving the original ground; hence, it is used as reference. In adopting this method, particular attention must be given to its investigation to confirm the applicability of the deep mixing method. If the ground under investigation contains gravel or stones, the application of the method is difficult. A small quantity of stones can be broken by a drilling machine. However, if a considerable amount is embedded, these stones must be dug up, removed, and replaced by soil and sand, which can be improved. The design of the method has considerable commonality with the deep mixing method for original ground improvement. The improved ground is considered a solidified body, and external stability (sliding, overturning, and bearing capacity) and internal stability (internal shear and edge toe pressure) must be verified. Note that the stability need not be considered with the sheet piles or other earth retaining walls; fundamentally, only the solidified body must be designed to ensure stability. A feature of this structure is that the weight of the main body made of cement-treated soil is lower than that of concrete. Moreover, the size of the main body must be larger than that of a normal gravity-type quay wall. The results of shaking table tests [5] indicate that the treated soil to be used as the quay wall body must have a certain width and founded on a layer that does not liquefy. Furthermore, treated solidified soils exposed to water and subjected to repeated dry and wet cycles are susceptible to damage, facilely degrading the cement-treated soil. The superstructure must be designed such that the treated soil near the ground surface does not deteriorate, and sheet piles must be attached to avoid seawater exposure.

Execution planning is particularly important for this quay wall structure. If the existing quay wall structure is of the sheet pile type, the agitator blades are difficult to insert from the ground surface because the sheet pile anchoring and tie rod are probably buried. If the existing quay wall has a margin of stability, execution is implemented while cutting some tie rods. However, with this margin, execution is implemented while installing supplementary temporary anchoring and tie rods. Because the ground in the vicinity of existing sheet piles is difficult to be improved by mechanical agitation (using agitator blades), it is improved by the jet grouting. If the existing quay wall structure is not of the sheet pile type or if this structure is applied to a newly constructed quay wall, an earth retaining wall is required to temporarily sustain the soil before improvement. Sheet piles constructed for preventing seawater exposure must also function as an earth retaining wall during execution. Otherwise, a temporary earth retaining wall must be constructed on the seaward side.

4.6.3 Reinforcing Levees Using Deep Mixing Method

During the 2011 Great East Japan Earthquake, an enormous tsunami surged onto the coast and broke through coastal levees. The overflowing water destroyed the back slopes of the levees, leading to collapse (Figure 4.13). After the earthquake, methods for covering the back slopes and reinforcing the toes of slopes have been proposed to prevent the collapse of levees after long periods of overflow. However, levees built on soft ground can deform or be destroyed by seismic motion, and their top may settle even before a tsunami strikes. To resolve this, the technique of improving the original ground near the bottom of the levee has been previously used. Furthermore, maintaining a top height that can resist considerable seismic motions and tsunami overflow while allowing the slope to fracture has been considered. Specifically, steel pipe sheet piles are cast in the middle of the levee, enabling the sheet piles to remain in place and maintain the top height even if the slope collapses due to seismic action or tsunami. This prevents tsunamis from entering the back land. After an earthquake, the levee can be reused by

Figure 4.13 Coastal levee destroyed by tsunami. (Courtesy of Dr. Yasuda (Kansai University).)

simply restoring the slope. This method was developed as a countermeasure against earthquakes and tsunamis. It is also effective in preventing levee collapse due to earthquakes and slope collapse due to tidal waves and wave overtopping.

Cement-treated soil prepared by the deep mixing methods can be applied as material to maintain the top height of the levee. Sheet piles are relatively expensive, and their embedment depth must be considerable for the sheet piles to be self-supported. Accordingly, the author has investigated a reinforcement technique in which a cement-treated soil wall is constructed on the support layer of the levee by deep mixing. The investigated technique maintains the top height even under seismic motion and overflow; its effectiveness has been clarified by centrifuge model tests [6, 7]. The method is characterised by the fact that the cement-treated soil walls on the support layer are self-supporting. Consequently, even if the slope collapses due to seismic motion or overflow, the back land is not swamped by the tsunami (Figure 4.14). For the cement-treated soil wall to be self-supporting, a block or grid of cement-treated soil of a certain width is provided. If the pore water pressure near the wall does not increase when a tsunami occurs, then the earth pressure acting on the wall is reduced. If the seismic resistance of the levee slope and ground inside the grid is increased and water penetration into the ground is prevented, a rational design of the cement-treated soil wall is possible.

The investigation of the foregoing technique is also similar to that of the deep mixing method for improving original ground. The fundamental design concept is also similar to that of the deep mixing method for original ground improvement. Hence, verifying the external stability (sliding, overturning, and bearing capacity) and internal stability (internal shear and edge toe pressure) is also necessary. If grid-type instead of block-type improvement

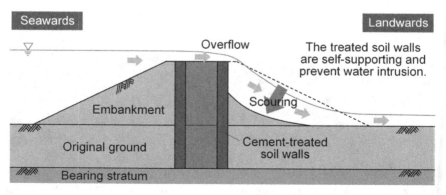

Figure 4.14 Concept of levee subject to tsunami overflow.

is used, checking the stability of the walls composed of grids against shear and bending failures is necessary. The difference of the foregoing from the case in which the original ground is improved by deep mixing is the external force. In the case of a tsunami, the water level conditions on the seaward and landward sides of the levee differ; these hydraulic external forces must be considered. Note that if the seaward slope remains intact when a tsunami strikes, water percolation is delayed; hence, the tsunami strike and high water pressure surge acting on the cement-treated soil wall have a time lag. If the tsunami recedes during this period, accounting for the external force due to the large water pressure is not necessary; hence, improved design can be achieved. However, if the surface slopes are scoured during the tsunami recession, the solidified body becomes unstable. In this case, examining whether countermeasures against surface slope scouring must be implemented is necessary. Such a measure can enable the width of the treated soil wall to withstand external forces when it is scoured, or design conditions can be set for the wall to withstand only the first thrust of pushing waves.

One of the execution problems with this method is ascertaining whether the implementation of deep mixing on the levee is possible. If the ground contains crushed stones or other hard materials, drilling is required; this may increase the execution cost. If the levee is not sufficiently wide, ground deformation (especially lateral displacement) may occur when the ground is improved by deep mixing. Moreover, if houses and other structures close to the levee are unable to tolerate deformation, low-displacement deep mixing machines or other methods must be used.

4.7 REINFORCEMENT WITH GEOSYNTHETICS

4.7.1 Overview

Slightly different from ground improvement, the use of geosynthetics reinforces or adds functionality to the ground. Geosynthetics (a compound term of 'geo' and 'synthetics') is the generic term for geotextiles, geomembranes, and geocomposites. The term geotextiles was first used to refer to petrochemical fibre materials for construction. Then, as the number of products increased, the term geosynthetics became a generic term. The concept of geosynthetics started during the BCE period with the idea of strengthening structures by placing plants in the soil. Subsequently, as the industry on ground improvement developed, such materials were replaced by petrochemical products. In recent years, a wide variety of these materials has been developed and used in many areas, not only for reinforcement but also for improving imperviousness, drainage, and protection, as shown by the classification in Figure 4.15.

Geosynthetics have been extensively used particularly in the construction of civil engineering structures on land. In contrast, its application in

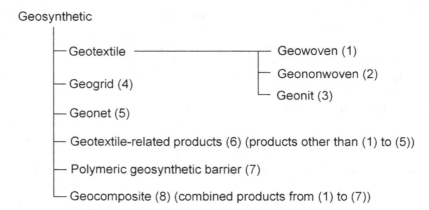

Figure 4.15 Classification of geosynthetics.

coastal works including onshore works is limited. This section describes geosynthetics for ground reinforcement and its functional use in coastal areas. Specifically, the following methods are introduced: laying of geosynthetics on shallow ground areas; using geosynthetics as a membrane to prevent sand and water from passing through; filling geosynthetics bags with sand and using them as large sandbags; and placing geosynthetics inside earth-retaining walls similar to land-based works.

4.7.2 Use in Shallow Layer Treatment

Geosynthetics is typically used as reinforcement when shallow treatment methods are implemented. As described in Chapter 3, if coastal structures are relatively small, only surface layers require improvement. One of the techniques in shallow treatment involves the use of bedding reinforcing material. In this regard, geosynthetics, especially geogrids and woven fabrics, are frequently used. In coastal reclamation, soft dredged material is typically used as backfill due to ease of procurement and cost considerations. However, dredged sediment is generally soft and may sink due to insufficient support when it is covered with soil of satisfactory quality. In the first place, the execution equipment used for levelling it may not be able to enter the site. To resolve this, fibre-based geogrids or woven fabrics can be laid to increase the bearing capacity of the ground.

Figure 4.16 shows the mechanism of increased bearing capacity when bedding materials are installed. As shown in the figure, the load is supported by the force from the ground below the bedding material, the tensile and friction forces of the material, the ground deformation suppression effect of the material, and the peripheral counterweight effect due to settlement.

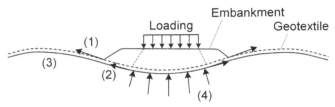

(1) Tensile force   (2) Friction force
(3) Ground deformation suppression effect
(4) Subgrade reaction including counterweight effect

Figure 4.16 Mechanism of increased bearing capacity due to underlying sheet.

The balance among these forces (as well as effects) depends on the ground, bedding material, and loading conditions.

The design of the foregoing involves determining the lining thickness and selecting the bedding material. The lining thickness depends on the required bearing capacity with the assumption that the load is distributed at 45–60° within the lining and acts on the bedding material. If the objective is to ensure trafficability, a layer thickness of 1.2–2.5 m must be used. The common practice is to divide the soil into layers instead of forming the cover at once. Moreover, a homogenous soil cover is advantageous. The first layer must be particularly thin. If the ground is overly soft, the thickness must be 20–30 cm, and if the soft soil has sufficient firmness for people to walk on, the thickness must be 50 cm. Next, the necessary specifications for the bedding material are determined using a calculation formula that considers the foregoing mechanism for increasing the bearing capacity. The bedding material satisfying these specifications must also be selected. In addition to tensile strength, the required performance includes elongation, frictional force with the soil, permeability, and durability. For example, geogrids, which are frequently used as bedding material, are factory-manufactured to a certain size (e.g., 1–2 m wide × 30–50 m long) and then bound together with fibre rope or tape when laid on site. The tensile strength of these tied sections must be accounted for in design.

4.7.3 Geosynthetics Used for Interception

The geosynthetics used in coastal areas that are expected to have with a filtration function are sand barrier sheets that are placed between backfill stones and soil behind quays and seawalls. As shown in Figure 4.17, the sand barrier sheets laid behind backfill stones (or rubble mounds) prevent sand from falling into the stone gaps due to waves and currents. Woven or non-woven factory-manufactured fabrics (10–20 m long) are laid on-site. Adjacent sheets

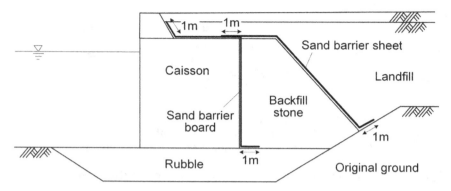

Figure 4.17 Layout of sand barrier sheets.

overlap by approximately 50 cm and are bound at predetermined intervals with fibre rope or tape. For non-woven fabrics, sheets that are at least 4 mm thick and can elongate to a minimum of 60% are used. For woven fabrics, the minimum thickness and elongation are 0.5 mm and 15%.

Sand barrier sheets, especially those with high imperviousness, are intended to prevent sand from escaping. A typical application of impervious sand barrier sheets is in offshore waste disposal sites. Worldwide, the standards for impermeable construction have become stricter for both onshore and offshore disposal. In Japan, for example, non-asphaltic impervious sheets must at least be 1.5 mm thick, whereas asphaltic sheets must have a minimum thickness of 3.0 mm. If the impervious sheet is expected to function solely as an impervious barrier, the use of double-layered sheeting is mandatory. Figure 4.18 shows an example of a double-layered impervious sheeting behind a seawall (caisson type) in a waste marine disposal site. To prevent the contact of the impervious sheet with the ground, the sheet must be sandwiched between additional protective materials (e.g., non-woven fabrics) to avoid damage. Furthermore, a protective layer is provided between the two layers of impervious sheets, as shown in the figure. Various types of impermeable sheets are available; however, those used in coastal areas must be submersible in seawater. After laying out the impervious sheets, the slopes over which the sheets are spread must be protected because the waste may not always be immediately set in place. For example, the cement-treated soil mentioned previously or slope protection works where mortar or concrete is poured into fabric bags may be used.

4.7.4 Geotextile Tubes

Beach erosion and cliff recession caused by waves are major problems in coastal areas. Various methods for resolving these problems are available.

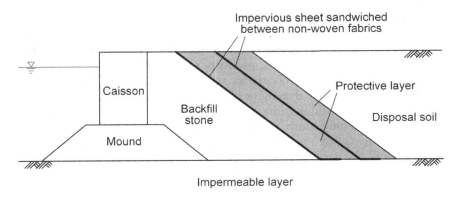

Figure 4.18 Impervious sheets for seawalls in waste marine disposal site.

These include approaches applied to the fluid external force side (such as weakening the waves) and resistance side (such as forming seawalls capable of withstanding wave forces). Geosynthetics are used in the latter method. Wave forces are resisted by filling large geosynthetic bags with local beach materials as well as beach nourishment materials and arranging them in rows. Filling the bags with sand enables rapid execution; hence, early deterrent effects can be expected. The bags require strength and durability. Hence, they have a double structure: a tubular bag material made of high-strength woven polypropylene as the base fabric and an outer sheet that protects the bag material from abrasion and ultraviolet radiation. In addition to protecting sandy beaches and cliffs, these sand-filled bags are used to form temporary structural seawalls or contain waste disposal. They were first used in the USA in the 1950s to prevent beach erosion in coastal areas. Today, the method is widely used throughout the world.

Slurry is poured into large bags and allowed to drain, thus filling them with sand. The diameter of one large bag, approximately 1.5–5.0 m, is selected according to the size of the structure. The protective surface can also be raised by piling up the sand-filled bags. To increase stability against waves, larger bags can be used. In examining this stability, the following actions are considered: forces from heavy machinery and pumps during execution, loads during service, self-weight, earth pressure from the ground behind the bags, wave forces, seismic forces, and impact forces (such as those due to sand and gravel motion). The instability caused by the foreshore being cut away over time and material degradation due to changes in ultraviolet light and moisture must also be considered. The items to be checked include the overall slope stability; deformation characteristics of the entire beach including suction; stability, strength, and durability of the large bags; and environmental impact.

The major concern with this method is durability. A base fabric with high initial strength is deemed to have excellent abrasion and weather resistance. In Japan, before implementing this method, several tests are conducted using products from various companies to understand their characteristics relative to the technique. Among these, products remain durable for at least ten years. However, if a notch occurs for some reason, damage may be rapid. If coating is applied, the durability of the product is presumed to increase. Hence, the long-term durability of the method is unknown depending on the product, and only future results can demonstrate its durability. In addition, durability is not a problem if the method is only used as a temporary measure until the sandiness of the beach is restored. Further, the packed sand can be removed and reused. Accordingly, the method can be deemed as advantageous.

4.7.5 Earth Pressure Reduction

The last application of geosynthetics in the coastal zone considered in this section is their use as earth retaining reinforcement. This technique, known as the reinforced earth wall method, is typically applied to onshore structures. Reinforcement is attached to the flat plate forming the wall surface and embedded into the ground behind the wall surface to form a vertical or near-vertical slope. The pioneering method that led to the development of reinforced earth wall construction was the taille-armé method. This technique, developed in France in 1963, employs a panel wall, a layer of steel reinforcement strips laid at regular intervals, and soil material. Various reinforced earth wall construction methods currently used are based on the tail-armé method supported with new knowledge (e.g., reinforcement material, type of wall, construction method, and design method).

In coastal areas, the reinforced earth wall is difficult to apply simply because compacting the ground underwater is not easy and the pull-out resistance of the reinforcement cannot be ensured due to erosion. An example of its application in coastal areas is the use of reinforcement attached to concrete blocks in small seawalls [8]. This application was considered possible because the seawalls were small, and land-based construction was possible. A research group including the author has investigated the use of reinforced earth wall construction in coastal areas without compacting the ground by employing sheet piles as a wall surface, attaching geogrids to the sheet piles, and placing crushed stones behind the sheet piles [9]. The technique has been found advantageous. The researchers observed that embedding the geogrids (polymeric materials in a regular grid pattern) with the crushed stone allows the stones to engage with the geogrids and provide pull-out resistance even if the ground is not dense. Furthermore, the strength of the geogrid exceeds that of mild steel and has been evaluated in terms of long-term durability, distinguishing it from other geotextile materials. Full-scale and centrifuge model tests have confirmed its effectiveness. Accordingly, it is considered to be a useful option in the future for sheet pile quay walls without piles.

In reinforced earth wall construction, shear resistance forces are generated between the reinforcement and soil, creating tension in the reinforcement and preventing the flat plate to which the reinforcement is attached from collapsing towards the front. Various design conditions and failure modes have been established to determine the reinforcement tension and arrangement to ensure the stability of the entire wall. In design, two main types of failure modes are considered. One is internal stability, which accounts for the internal failure of the reinforced area (rupture of the reinforcement or its pull-out from the soil and rupture of the reinforcement in the slabs forming the wall surface). The other is external stability, which considers the stability of the entire reinforced soil structure (sliding, overturning, bearing failure, and sliding through the exterior of the structure). By examining both modes, the integration of areas to be reinforced is possible. This ensures stability against the external forces acting on the area and of the entire system, including the ground supporting the area. A similar design is considered possible for this method. However, in considering the external forces, accounting for the water level conditions specific to coastal areas is necessary.

References

1. Coastal Development Institute of Technology. 2003. *The Premixing Method: Principle, Design and Construction.* CRC Press.
2. Coastal Development Institute of Technology. 2017. *Technical Manual on the Use of Calcia Improved Soil at Harbours, Airports and Seawall.* Coastal Development Institute of Technology (in Japanese).
3. Kitazume, M. 2020. *The Pneumatic Flow Mixing Method.* CRC Press.
4. Tsuchida, T. and Egashira, K. 2004. *The Lightweight Treated Soil Method: New Geomaterials for Soft Ground Engineering in Coastal Areas.* CRC Press.
5. Takahashi, H., Fukawa, H., Asada, H., Nguyen, B., and Takeuchi, H. 2022. Seismic behaviour and stability assessment method of quay walls constructed with cement-treated soil. *Journal of Japan Society of Civil Engineers, Ser. B3 (Ocean Engineering)* 78(2): I_589–I_594 (in Japanese).
6. Takahashi, H. 2022. Tsunami overflow tests of levee reinforced by deep mixing method in centrifuge. Proceedings of the 20th International Conference on Soil Mechanics and Geotechnical Engineering, 1211–1215.
7. Takahashi, H. 2022. Dynamic model tests of levees reinforced by cement deep mixing method. Proceedings of the 10th International Conference on Physical Modelling in Geotechnics, 939–942.

8. PIANC. 2011. The application of geosynthetics in waterfront areas. PIANC Report, No. 113, pp. 52–55.

9. Takahashi, H., Morikawa, Y., Mizutani, T., Ikeno, K., Tanaka, T., Mizutani, S., Miyoshi, T., and Hayashi, K. 2018. Technological development of sheet-pile quaywall reinforced by geogrids. *Report of the Port and Airport Research Institute* 57(1): 36–81 (in Japanese).

Chapter 5

Improvement of Contaminated Soil

This chapter describes ground improvement as a countermeasure for soil contamination prevention and environmental protection. Waste containing hazardous substances may have been used as landfill material in coastal areas. Another possibility is that the soil may have been contaminated by economic activities in these areas. In this case, the ground can be improved to prevent hazardous substances from leaking into the surrounding ground or sea. The main purpose of soft ground improvement described thus far is to increase ground strength. However, herein, the goal is to prevent the leakage of hazardous substances into surrounding areas. The foregoing is regarded as a type of ground improvement, and it is elucidated in this chapter.

5.1 MEASURES AGAINST CONTAMINATED SOIL

Soil pollution caused by human activities has threatened human health and led to the destruction of ecosystems. In some cases, because economic development has been prioritised, environmental protection measures have not been implemented. Occasionally, soil has been used without recognising that it is contaminated by hazardous substances. Considerable amounts of waste from industry and human activities are typically discharged into large coastal areas. Some harmful substances that are of natural origin are also introduced to soil. Typical hazardous substances include volatile organic compounds (VOCs), heavy metals, and pesticides. Although efforts continue to be exerted worldwide to reduce the use of hazardous substances and replace them with alternative materials, measures against contaminated soils remain limited.

As a measure against contaminated soil, the conduct of an initial investigation to determine whether hazardous substances are present in the ground is necessary. The objective is to ascertain the type and concentration of hazardous substances in the ground, their diffusion into surrounding areas, and other contamination conditions. Two types of investigations can be conducted: statutory surveys (based on national laws and municipal regulations) and voluntary surveys (voluntarily implemented by operators).

DOI: 10.1201/9781003267119-5

Statutory surveys are conducted when the operations of facilities that have previously used hazardous substances are discontinued, when the handling of hazardous substances is halted, or when large-scale land alterations are implemented. These surveys are also performed when the administration deems that polluted soil in certain areas poses a health hazard. In some cases, operators conduct voluntary surveys to supplement statutory surveys. Some soil contamination problems are caused by natural sources; however, many of these sources, such as the leakage of hazardous substances or improper waste disposal, are man-made. In such cases, the possibility of multiple hazardous substance contamination or contamination by substances other than those specified in the law exists. Accordingly, soil investigations must be implemented considering the foregoing situations.

Contaminated soil measures include the prevention of the direct ingestion of contaminated soil or groundwater. These measures can be broadly classified as follows:

1. Exposure control: potential human contact with contaminated soil and groundwater is regulated.
2. Interception of exposure pathways: the migration of contaminated soil or hazardous substances in contaminated soil is controlled.
3. Removal of hazardous substances: harmful substances in contaminated soil are extracted or decomposed.

The possible countermeasures include the prohibition of entry sites with contaminated soil; filling, covering, and paving contaminated ground; water quality monitoring; containment and insolubilisation of contaminated soil; and removal and purification of toxic substances from contaminated soil. Among the foregoing, the ideal approach is to excavate, detoxify, and return the contaminated soil to its original location. However, this measure has considerable cost, and certain substances are difficult to detoxify. The excavation and detoxification of all contaminated soils are not feasible. First, the degree of toxicity of different hazardous substances varies, and whether they can cause health damage considerably depends on their concentration. Therefore, depending on the type and concentration of hazardous substances and their elution into groundwater and surrounding environment (e.g., surrounding facilities), improving the ground in situ to detoxify hazardous substances and preventing their leakage may be possible.

The methods for dealing with contaminated soil using ground improvement fall under the 'containment and insolubilisation of contaminated soil' and 'removal and purification of toxic substances from contaminated soil' countermeasure methods. Various specific methods are available; however, the main techniques include cement treatment, vacuum extraction, flushing, and in situ remediation. In the cement treatment method, the ground around the contaminated soil is solidified or the contaminated soil itself is solidified to prevent hazardous substances from leaking into the surrounding areas. Vacuum extraction is a method for extracting contaminated gases in which an air permeation well is built on the ground and then a vacuum pump is employed to remove the contaminated gases. In impermeable soils, such as clay, lime is first mixed to evaporate water and increase permeability before extraction. In the flushing method, water, surfactants, or solvents are injected into the ground, and the groundwater containing hazardous substances is collected from the pumping well. In the in situ remediation method, a ground mixer is used to blend remediation materials, microorganisms, plants, and other organisms into the ground to purify the contaminated soil. In addition to these methods, many other techniques have been proposed to manage contaminated soils. Hazardous substances are classified over a wide range, and the applicable remediation methods differ depending on the type of harmful substance. Hazardous substances include VOCs, benzene, oil, heavy metals, dioxins, and polychlorinated biphenyls (PCBs). Benzene, oil, and VOCs are highly mobile in the ground; they can be chemically or biologically decomposed. Accordingly, many of the foregoing measures can be used. In contrast, the mobility of heavy metals is low, and decomposing them is difficult; consequently, countermeasures are limited to several methods such as the cement treatment method. The countermeasures for dioxins and PCBs are also limited to the application of cement treatment and high-temperature pyrolysis methods. Hence, checking the manuals and other information on each of the countermeasures to confirm their suitability is important.

This chapter focuses on a relatively versatile cement treatment method as a countermeasure. A conceptual diagram is presented in Figure 5.1, and an overview is provided as follows:

1. The ground containing contaminated soil may be controlled by providing cover or embankment if the degree of harm is low and the hazardous substance does not spread to the surrounding area. In this case, cement-treated soil can be used to prevent damage to the soil cover or embankments. Cement treatment can reduce the permeability of soil and the possibility of exposure to hazardous substances.
2. If the spread of hazardous substances to the surrounding area is possible, a vertical impermeable barrier or wall can be constructed around the contaminated soil using cement treatment. The low permeability of cement-treated soil prevents the ingress and egress of groundwater between the contaminated soil and surrounding area, thus providing a measure of protection against contaminated soil.
3. The contaminated soil containing hazardous substances can also be cemented to prevent hazardous substances from leaking; this is called insolubilisation. Similar to Method 2, the low permeability of cement-treated soil prevents the entry and exit of groundwater between the contaminated soil and surrounding area.

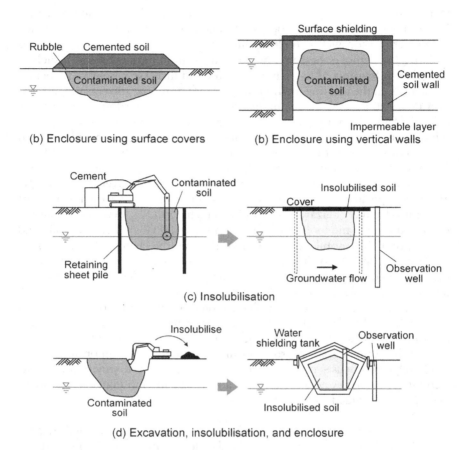

Figure 5.1 Conceptual diagram of cement solidification process.

4. If the degree of threat is high, in situ treatment is not advisable. Upon excavation and insolubilisation, the treated soil is returned to the ground and then enclosed and contained by impermeable structures.

Methods 2 and 3 are presented in this book. Method 3 is described in the next section, and Method 2 is elaborated thereafter.

5.2 INSOLUBILISATION OF CONTAMINATED SOIL BY CEMENT TREATMENT

5.2.1 Overview

This section focuses on the insolubilisation treatment of contaminated soil. It describes the insolubilisation mechanism and the effect of cement, which is an insolubilising material, on heavy metals and other substances. The use of cement as an insoluble material is advantageous because it is relatively inexpensive. Moreover, existing ground improvement equipment can be utilised, and ground strength can be increased. Shallow and deep mixing methods can also be used, and their insolubilisation mechanism and effectiveness have already been confirmed. These methods are applied not only to heavy metals but also to waste materials, such as incinerator ash. They are used worldwide for contaminated soil control. However, the use of cement requires some important considerations, such as the re-elution of heavy metals due to high pH and the elution of hexavalent chromium. Hence, the effectiveness of cement must be confirmed in advance through laboratory mixing tests. In addition, lower limits are set for the amount of cement added (e.g., 150 kg/m^3 or higher) and the strength of the treated soil (e.g., 1000 kN/m$_2$ or higher).

During cement treatment, the insolubilisation of heavy metals and other substances involves three mechanisms:

1. Fixation by formation of insoluble substances
   In general, metal ions, including heavy metals, are less soluble in alkaline solutions because of the formation of hydroxide precipitates, which greatly reduce their solubility. The cement reacts with water to form calcium hydroxide, which is alkaline. The reaction of hydroxide ions with heavy metals and other substances produces hydroxide precipitates, which are deposited and remain in improved soil. However, Pb, Zn, and other metals are amphoteric metals, that is, their solubility increases under both acidic and alkaline pH conditions, and insolubilisation may not be sufficient if the pH range exceeds the appropriate range.

2. Substitutional solid solution and surface adsorption by hydration products
   Cement hydrates form various hydrates via different reactions. Heavy metals are fixed and insoluble owing to the displacement solubilisation in the hydrates or adsorption on the hydrate surfaces. In particular, ettringite and monosulphate, which are easily produced in the cementation of soil, can immobilise heavy metals, such as hexavalent chromium and arsenic, by the displacement of solid solutions in their crystal structures or by physisorption on their surfaces.

3. Containment by densification of hardened body structure
   The strength of cement-treated soils increases with the age of the soil after mixing the cement. This is because the hydrates generated by the hydration reaction of the cement fill the voids in the soil and densify the structure as the soil age increases. This structure densification not only results in strength development but also in the physical containment of heavy metals and other substances within the improved soil. It also reduces permeability, thus reducing the leakage of heavy metals and other substances to the surrounding areas.

Insolubilisation by cement treatment is characterised by a combination of the foregoing mechanisms, enabling the stable solidification and insolubilisation of various heavy metals and other substances.

To ensure that the solidification and insolubilisation effects of cement are sufficiently effective, the following points must be considered:

1. The hydration reaction proceeds unimpeded, and improvement results in sufficient physical strength.
2. Treatments (e.g., covering) to prevent the action of external factors, such as carbonation and drying, must be applied to the improved soil.
3. Considering the nature of hazardous substances, some substances, such as Pb, tend to increase their solubility when the atmosphere has a high pH.
4. Ensure that the surrounding environment is not detrimentally affected during or after improvement execution (dust, noise, alkalis, etc.).
5. The effectiveness of solidification and insolubilisation of contaminated soil and industrial waste is affected by the physicochemical properties of the target hazardous substance and soil. Therefore, the effectiveness of the measures must be confirmed by laboratory mixing tests.

If the foregoing points are considered, the physical and chemical stabilities of the improved soil after the solidification and insolubilisation of the contaminated soil can be maintained for a long period.

The insolubilisation mechanism and its effect on heavy metals are described in this section. The use of the cement solidification process to insolubilise benzene, oil, dioxins, and PCBs continues to gain interest. It is also employed for soil odour and acidity control and has a wide range of applications. Although a description of the mechanisms of the treatment methods is omitted, the following laboratory mixing tests and points of caution are common to many of them and must be referred to.

5.2.2 Investigation and Design

Before implementing measures to solidify and insolubilise contaminated soil by cement treatment, its effectiveness must be investigated through laboratory mixing tests. Elution tests, as part of the laboratory mixing tests, are performed using the prepared cement-treated soil. Various methods, including batch tests, column flow tests, tank leaching tests, and pH dependence tests, have been employed. In the batch test, soil samples are air-dried and crushed to small particles (e.g., 2 mm or less), placed in a solvent, and shaken to determine the eluted components. This test method is simple and has been used in many countries. In the column leaching test, a soil sample is filled into a column, and pure water is poured from the lower end of the column to saturate the soil before it is leached. The water that flows out is analysed to determine the presence and quantity of hazardous substances. Compared with the batch test, this test is implemented under conditions similar to the actual environment; however, it requires more work. In the tank leaching test, soil samples are immersed in the bulk solvent and left to stand for a certain period of time (e.g., 1, 28, or 64 days); then, the solvent components are measured. In another test, the pH dependence test, a curve of the chemical equilibrium concentration of the component of interest (pH-dependence curve) is obtained with pH as the abscissa. Because the leaching behaviour of heavy metals is highly dependent on pH, the amount of toxic substances leached can be investigated by varying the pH level. Because the type of toxic substance, the purpose of the countermeasure, and the test methods required in different countries and regions differ, an investigation of the required tests and methods is necessary before implementation.

Figure 5.2 shows the results of the batch tests conducted in accordance with Notification No. 46 of the Environmental Agency of Japan on soils prepared from cement-treated soil. The addition of a cementitious solidifier (special cement for soil) reduces the amount of leaching; its effect increases with the amount of cementitious solidifier added and age of the material. As shown in Figure 5.2(b), the insolubilising effect of Pb decreases as the amount of the solidifier added is increased; leaching tests are important as such characteristic may occur. In the batch test, the soil samples are air-dried and pre-treated by particle sizing; the concentration of the trace constituents eluted into the solvent is determined by shaking. The surface of cement-treated soil is typically unexposed by paving or covering. Because the contact of the surface with the surrounding environment is limited, the surface is less susceptible to drying and carbonation. Consequently, the actual impact of the foregoing on the surrounding environment is lesser compared to that when the sample is crushed and air-dried. The tank leaching method is one of the tests reproducing the actual leaching environment conditions. The sample is immersed in water with a solvent having a solid–liquid ratio of 10. The concentration of trace constituents eluted into the water is evaluated after 24 h of immersion. In some cases, leaching properties are investigated using this technique. In addition, simulations, such as advection–dispersion analysis, can be used to evaluate the elution behaviour after soil solidification and insolubilisation.

The method for determining whether the formulation controlling the elution of a substance is below the standard value is described as follows. Laboratory mixing tests for examining the conditions and effectiveness of countermeasures in advance are also referred to as treatability tests. The objective of these tests is to determine the amount of additives that can be controlled below the standard value. Because cement-treated soils exhibit high alkalinity, some substances may increase the environmental risk and must therefore be thoroughly examined by laboratory mixing tests before

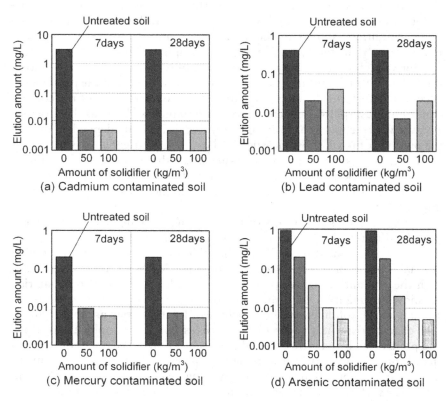

Figure 5.2 Example of elution reduction by cement treatment.(Adapted from Ref. [1]).

use. The procedure for the laboratory mixing tests is similar to that for strength expectations, as described in Chapters 3 and 4. However, a different test item is added: leaching test. The leaching test is based on the use of samples obtained during mixing. Because the amount of sample required for the leaching test is generally approximately 400–500 g, the specimens employed in the compressive strength test may be used. The salient points to be considered in the laboratory mixing tests of contaminated soil include the following. First, the condition under which the soil sample was collected must be determined, accounting for the contamination situation and soil properties. The soil sample is sealed and stored in a cool dark place to prevent deterioration due to drying or oxidation. If gases are generated during the mixing of the soil and solidifier, the experiment must be immediately halted, the test area must be ventilated, and the soil sample must be properly disposed. Select test tools (storage containers, mixing containers, formwork, etc.) made of materials that are not degraded by the assumed hazardous substances. Appropriate protective equipment, such as gloves, safety glasses, and masks, must be used during the tests. Proper soil disposal and washing using water after the leaching tests are necessary. The age of the cement-treated soil and additional test parameters (e.g., content, strength, pH, redox potential, and electrical conductivity) must be determined in consultation with relevant personnel.

As described in Chapters 3 and 4, when the strength of the cement-treated soil is expected, the amount of solidifier added is determined using the relationship between the amount of solidifier and developed strength given by the laboratory mixing tests multiplied by a factor (e.g., in situ–laboratory strength ratio). In contrast, in the insolubilisation technique, the existing knowledge on the in situ–laboratory insolubilisation performance ratio is limited. Presently, no method has been established for setting the amount of in situ addition from laboratory mixing tests. Therefore, various methods and concepts for setting the in situ additive amount are used in actual execution. The following technique is typically adopted. First, the amount of the solidifier that satisfies the standard value is determined. Then, this amount is multiplied by a premium rate. Finally, the resulting value is the amount of solidifier to be added in situ. Figure 5.3 shows a conceptual diagram for determining the amount of solidifier to be added. The relationship between the amounts of the solidifier and leached material is determined in the leaching test, and the amount of the solidifier required to remain below the standard value is determined. The amount of solidifier is then multiplied by a premium factor (e.g., 30%–50%). The existing surcharge

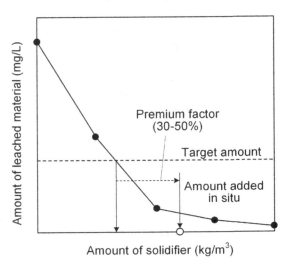

Figure 5.3 Conceptual diagram for determining amount of binder added.

is a coefficient that determines the amount to be added in situ considering the mixing accuracy based on strength properties and ignoring the leaching performance. Therefore, the method for establishing the solidifier addition amount and safety factor must be selected in consultation with the parties concerned in the actual execution. Regardless of the method used, verifying the adequacy of the amount of solidifier to be added in situ through test execution is necessary to confirm that the insolubilising effect satisfies the standard value.

5.2.3 Execution

The execution of solidification and insolubilisation is similar to that of ground improvement using the normal cement treatment method. However, mixing greatly affects the insolubilisation effect. Hence, selecting the execution machinery and method is necessary after thoroughly considering the soil type and extent of contamination, the amount of solidifier added, the execution period, and other factors. Another point to be noted during solidification and insolubilisation is that workers may inhale dust and toxic gases containing hazardous substances. Thus, the use of protective equipment, such as safety glasses and protective masks, is appropriate. In addition to dust, the diffusion of toxic substances into the surrounding area through adhesion to heavy equipment, such as execution machinery and transport vehicles, must be given attention. In particular, when facilities, such as residences or drinking wells, are found in the vicinity of the site, utmost precaution must be observed. The erection of temporary structures to prevent dispersal and the use of impermeable sheets and sheet piles to prevent groundwater seepage are effective. Depending on the situation, dust concentrations at the site boundary and concentrations of hazardous substances in the groundwater may be monitored.

The execution management during ground improvement follows the methods used for normal cement treatment. Because the management methods vary according to the execution method, the execution management of corresponding methods must be referred to. For example, if the deep mixing method is used, the water–cement ratio of the cement slurry, discharge rate, speed of execution, and number of blade cuts are specified. For quality control, the leaching tests of cement-treated soil must be performed using fixed quantities. In situ insolubilisation tests are typically implemented by sampling the soil from a depth of 1 m to the depth of the insoluble contaminated soil at a rate of 1 point per 100 m^2 area of the site. The results of the elution tests are not immediately available; hence, in some cases, a simple analysis is used for on-site management. When a simple analysis is used, ensuring its accuracy and correlation with the original elution test is necessary considering the contamination status (e.g., hazardous substances, valence, and form of existence).

An important aspect of contaminated soil remediation is monitoring after execution. Monitoring involves periodic investigations of the surrounding groundwater and other contaminants to ensure that the implemented measures are effective. Monitoring is sometimes required by law. For example, in Japan, monitoring is stipulated in the Soil Contamination Countermeasures Law, whereby observation wells are set up at one or more points downstream of the area to be treated. Groundwater is periodically investigated at a rate of at least four times a year. For a period of two years, the target groundwater concentration must not be exceeded; this determines whether insoluble treatment has been properly implemented. In addition to the concentration of hazardous substances, the pH, redox potential, and electrical conductivity are measured in the observation wells to check whether the values of these parameters remain unchanged and to determine the persistence of the insolubilisation effect.

5.3 ENCLOSURE OF CONTAMINATED GROUND USING CEMENT-TREATED SOIL

5.3.1 Overview

Another countermeasure against contaminated soil is to construct a vertical impermeable barrier wall using cement-treated soil as an enclosure around the soil. A typical civil structure in coastal areas where this countermeasure method is applied is a maritime waste disposal site where the sea is reclaimed using waste material. Coastal areas have long been used as disposal sites for the general and industrial waste generated by humans. In the past, illegal dumping and unsystematic management have led to environmental destruction; however, nowadays, the waste disposal in these areas is accepted with proper management. Once the repository has reached its capacity, it is used as land in port areas. With appropriate design, execution, and management, the construction of more waste disposal sites in coastal areas is highly possible in the future. To prevent the leakage of hazardous substances, a perimeter seawall may be constructed in a maritime waste disposal site. If an impermeable layer (e.g., 5 m thick with a hydraulic conductivity of less than 1×10^{-5} cm/s) is present in the underlying ground below the disposal site, a vertical impermeable barrier can be constructed adjacent to the seawall, and the waste can be disposed within the barrier. Similar to vertical impermeable barriers built on the ground, cement-treated soil walls are built in addition to these barriers using chemical grouting and impermeable sheet piles. If no impermeable layer is present in the underlying ground, a fully impermeable barrier must be provided before the waste is disposed to prevent hazardous substances from flowing out to the side or downwards. The waste accepted in a repository is generally limited to less hazardous waste; the area is known as a controlled waste disposal site. Highly hazardous wastes must first be stabilised to reduce their hazardousness before their disposal into a site. Strictly controlled waste disposal sites, which require a higher level of management than controlled waste repositories (i.e., completely sealed sites), are not usually constructed in coastal areas due to management difficulties.

Recently, cement-treated soil has been widely used as a vertical impervious barrier in seawalls owing to its long-term durability, reliability, and economic efficiency. The principle of preventing harmful substances from leaking out through the walls of cement-treated soil is the same as the insolubilisation principle described in the previous section. Reducing permeability by densifying the treated soil is particularly important. Although diffusion and advection cannot be completely stopped because water remains in the pore spaces of the cement-treated soil, their speed is extremely slow, and virtually no leakage of hazardous substances occurs at concentrations affecting the environment. In addition, the flow of hazardous substances is expected to be confined within the cement-treated soil due to the replacement solubilisation by hydration products and surface adsorption. Furthermore, heavy metals may be immobilised by the formation of insoluble substances. Note that the impermeable underlying soil below the disposal site is composed of clay, and the overburden load acting on the clay deforms the underlying soil when waste is disposed. This deformation displaces the seawall; in turn, the cement-treated soil wall is stressed and displaced. Ensuring that the cement-treated soil can conform to the deformation or that the strength and thickness of the wall of the cement-treated soil are sufficient is important. Thus, despite the deformation of the cement-treated soil, failure does not ensue. Because the treated soil is brittle, mixing rubber chips or fibres with the soil is considered to improve soil's ability to conform to deformation.

5.3.2 Investigation and Design

In contrast to the solidification and insolubilisation measures for contaminated soil by cement treatment, when the contaminated soil is enclosed by a seawall, the investigation and design of ground improvement in a structure are necessary. For the investigation prior to design, laboratory mixing tests are conducted to determine the type and amount of solidifier to be added to obtain the required strength, permeability, and environmental impact. Similar to the normal cement treatment method described in Chapter 3, these tests are in addition to the in situ ground investigation for determining the depth and thickness of the layer to be improved. A characteristic feature of the laboratory mixing test is the determination of permeability. This is because permeability indicates whether no hazardous substances leak into the surrounding area. However, the permeability test of cement-treated soils is difficult to implement. In this regard, the American Society for Testing and Materials specifies the use of a soft-walled apparatus

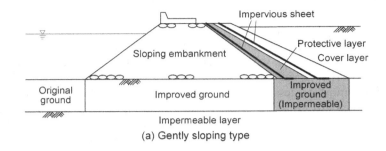

(a) Gently sloping type

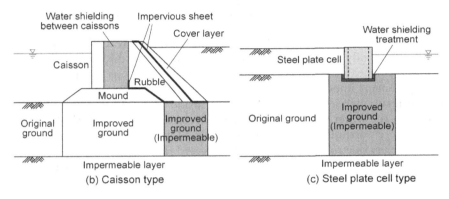

(b) Caisson type

(c) Steel plate cell type

Figure 5.4 Cross-sections of seawalls at waste disposal sites.

as an option. If the normal falling-head permeability test is not suitable, the easiest method is to use the apparatus of a triaxial compression test. The specimen is placed in a triaxial cell, stress is applied to the sidewall through the membrane, and permeation is implemented under this condition such that no water leakage occurs on the sidewall.

Examples of cross-sections of seawalls at a maritime waste disposal site using cement-treated soil walls as vertical impermeable barriers are shown in Figure 5.4. Cement-treated soil walls are used as vertical impermeable barriers in the foundations of gently sloping and gravity-type seawalls. In the examples shown in Figure 5.4(a) and (b), impervious sheets are installed in the backfill of the superstructure, and the cement-treated soil wall is built in the lower part of the backfill such that the impervious wall is continuous. In the example shown in Figure 5.4(c), the superstructure is impervious, and the ground immediately below the superstructure is improved by the cement treatment. In this case, the bearing capacity of the superstructure is also provided by the cement-treated soil. For treated soil to improve the stability of the structure, the stability of the treated soil itself must be ensured, as described in Chapter 3, which provides more information on this approach. The cross-sections shown are only examples, and the appropriate structural type must be selected according to site conditions.

One of the design considerations that must be satisfied is impermeability. For example, in Japan, when cement-treated soil is used as a vertical impermeable barrier, its thickness must at least be 50 cm, and the hydraulic conductivity must be less than 1×10^{-6} cm/s. The method for determining the permeability using laboratory mixing tests is described above, and the amount of cement to be added is set such that the permeability achieved is below the standard value. Advection and diffusion analyses can also be conducted to determine impermeability. Ensuring that water and hazardous substances are contained within the seawall during the required service life of the structure (e.g., 50–100 years) is necessary. In addition, the fail-safe concept along with back-up functions and water-level management is typically introduced to the seawalls of waste disposal sites. Even if the impermeable barriers are damaged, other functions can be designed to supplement impermeability. The back-up function provides a separate impermeable barrier, whereas the water inside the seawall is managed by lowering the water level such that it is advected into the seawall. In countries where earthquakes are common (e.g., Japan), damage to impermeable barriers due to seismic forces that exceed the expected magnitude has a non-zero possibility. In such cases, a fail-safe approach is particularly important.

5.3.3 Execution

The execution procedure for the methods used in the typical cement treatment is described in Chapter 3. One of the considerations in using cement-treated soil as the enclosure of contaminated soil is its impermeability. For example, in the deep mixing method where the ground is solidified by agitator blades, an improvement wall is constructed by overlapping the improvement piles formed via a single driving operation; if this overlap is insufficient, the impermeability cannot be maintained. Hence, the accuracy of the pile overlap and the verticality of the improvement in pile driving must be ensured. Execution management and quality control also follow the methods used for the normal cement treatment. If the deep mixing method is used, execution management considering the water–cement ratio of the cement slurry, discharge rate, execution speed, number of blade cuttings, and quality control (such as check boring after solidification) is necessary. Because the deformation of the seawall can cause the failure of the cement-treated soil, monitoring the displacement of the seawall during the construction of the superstructure and reclamation is necessary to ensure that deformation does not exceed the assumed value in the design.

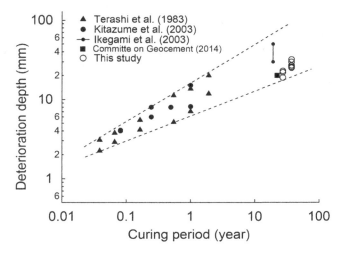

Figure 5.5 Relationship between number of years and degradation depth from exposed surface. (Adapted from Ref. [2]).

The principle of the foregoing similarly applies to the cement solidification and insolubilisation measures for contaminated soil. However, the long-term deterioration of cement-treated soil must be considered when enclosing contaminated soils with a cement-treated soil wall. Different from concrete, pore water remains in the treated soil and is connected to the surrounding soil through pore water. This leads to a long-term decrease in the concentration of calcium ions in the treated soil, resulting in the breakdown of cement hydrates. Although the rate of deterioration of cement-treated soils is low, the results of a study on deep mixing treatment methods show that the relationship between the number of years of exposure and depth of deterioration from the exposed surface is as shown in Figure 5.5. The figure indicates that the relationship between the number of years and degradation depth on both logarithmic graphs is linear. However, the relationship is not proportional, and the increment of degradation depth decreases with time. Accordingly, the rate of degradation during the first year can be inferred as approximately 10 mm. Thereafter, the degradation rate decreases and the degradation depth remains at approximately 20–100 mm after 100 years. If a 50-cm-thick vertical impermeable barrier is installed, a thickness of at least 30 cm is anticipated to remain after 100 years, even if degradation occurs on the exposed surfaces on both sides. However, the degradation rate may vary depending on the ground and execution conditions. The degradation characteristics of cement-treated soil must be clarified in the future.

References

1. Japan Cement Association. 2021. *Soil Improvement Manual for Cement-based Stabilizer (5th ed.)*. Gihodo Shuppan (in Japanese).
2. Takahashi, H., Morikawa, Y., Fujii, N., and Kitazume, M. 2018. Thirty-seven-year investigation of quicklime-treated soil produced by deep mixing method. *Ground Improvement* 171(3): 135–147.

Chapter 6

Future of Ground Improvement

The author's perspectives on future ground improvement in coastal areas are presented in this final chapter. Over the last 50 years, ground improvement technology has remarkably improved, and various improvement principles and techniques have been developed and implemented. Existing ground improvement methods have been enhanced and applied. This trend is expected to continue in the future, and new challenges will continue to be confronted. The prediction of everything regarding the future of ground improvement is impossible. Nevertheless, in view of current trends, the aspects that are anticipated to be considered in the field of ground improvement are worth summarising.

6.1 DEVELOPMENT OF NEW GROUND IMPROVEMENT METHODS

As indicated by the present trend, the development of new ground improvement methods and the advancement and application of existing methods are expected to continue. For example, when insufficient strength or consolidation settlement of soft cohesive soils was a problem, the replacement method and techniques for forming improved piles using sand or gravel were developed. When liquefaction during earthquakes became a problem, compaction methods and grid-type improvement approaches were developed. In the future, ground improvement methods will be devised, and the application of existing methods will continue to satisfy new requisites that may arise. For example, in coastal areas, flood damage, such as that caused by tsunamis and tidal waves, requires resolution. In the past, countermeasures against levee failure have been considered. However, currently, sea levels are rising owing to climate change. Moreover, the extent of damage caused by hurricanes, typhoons, and cyclones has increased. Because civil engineering structures in coastal areas are subject to external hydraulic forces, the elucidation of the combined phenomena occurring on water and ground and the implementation of countermeasures that utilise ground improvement

DOI: 10.1201/9781003267119-6

are necessary. Therefore, special requisites must also be satisfied. For instance, the demand for ground improvement work in narrow areas, areas with height restrictions (e.g., near an airport or under a bridge), and in the vicinity of existing structures continues to increase each year. The development of small execution machines that can easily improve the ground directly under existing structures is also anticipated.

One of the key requirements is to increase cost-effectiveness for ground improvement to be more economical than it presently is. To ensure ground improvement effectiveness, the volume and ratio of improvement can be reduced, execution equipment can be modified, and the costs and time required for each ground improvement process can be reviewed. For example, in the sand compaction pile (SCP) method, the volume and ratio of improvement are reduced if the site can tolerate a certain degree of displacement. Reducing the volume and ratio of improvement increases displacement and reduces stability; therefore, the accurate prediction of deformation and assessment of stability are required. In the future, improved prediction and evaluation accuracies may lead to greater economic efficiency. In addition, enhancing the execution accuracy of ground improvement and the efficiency of execution machines may reduce the volume and ratio of improvement and consequently increase economic efficiency.

Technological improvements in the existing ground improvement methods are expected to continue. Even with methods that have already been developed and implemented, improvement effectiveness may not be fully predictable, or the improvement mechanism may not be clear. In such cases, fundamental research and technological advancements will progress. For example, in the case of compaction methods, the degree of influence of the area of compaction is difficult to quantify, and the estimation equations specified in the design are typically not applied. Enhancing the accuracy of estimation equations through research and field demonstrations is also important; hence, works on the foregoing are expected to continue in the future.

New materials can also be used for ground improvement. In the case of cement treatment methods, the development of new cements with high improvement potential is expected. Further, techniques for mixing by-products, such as air bubbles, tire chips, and ash, will continue to advance. Relatively small amounts of other materials can also be added and utilised. For example, the author [1] has developed a technique in which cellulose nanofibres (CNFs) are added to enhance the performance of cement-treated soil using an additive. The addition of CNFs has been found to improve the homogeneity of cement mixes, increase initial mix strength, and exhibit thixotropy. Hence, the development and use of other additive materials may solve the problems encountered in the use of cement-treated soils. In addition to cement treatment methods, the advancement of ground improvement technologies using new materials is expected (e.g., the use of sand fluidised with polymeric materials in place of mortar or bacteria to solidify soil).

6.2 EXPANSION OF APPLICATION AREAS FOR EXISTING TECHNOLOGIES

The use of existing ground improvement technologies in areas and applications where they have not been previously considered is another possible direction in the future development of ground improvement. For example, the execution of cement-treated soil walls using deep mixing as described in Chapter 4 may be applied to quay wall bodies and core materials for coastal levees. As shown by these examples, the application of cement treatment may be facilely expanded. At the time of its development, cement treatment was mainly used for improving the original ground to support embankments and superstructures. However, currently, it is used in a wide range of applications, including liquefaction countermeasures, increasing the passive resistance of sheet piles, and reducing the main earth pressure on retaining structures. The use of sheet piles instead of steel or concrete piles as foundations for bridges and buildings has become feasible. Therefore, the range of applications is expected to increase.

Consider the following expansion of ground improvement applications. In the past, SCP and compaction grouting methods for compacting sandy soils are used as liquefaction countermeasures. Currently, to prevent liquefaction, the chemical grouting method is applied to replace sand voids with gel. Hence, the use of existing technologies in other applications can be expected.

6.3 USE OF INFORMATION TECHNOLOGY

Section 6.2 describes the use of information technology in the execution of ground improvement. Specifically, it includes the introduction of information and communication technology (ICT) for the implementation and automation of the execution process. The display of information during execution (e.g., agitator blade tip depth, penetration and withdrawal speeds, rotational speed, and cement slurry flow rate) in an animated form is possible on tablet terminals located outside the execution machine. Augmented reality technology can be used to superimpose the information regarding buried objects and their location on a shooting screen. The implementation of other measures to improve the efficiency and sophistication of ground improvement is also ongoing. For instance, building information modelling (BIM) and the sharing of information on 3D models that integrate the project management of ground improvement among the parties concerned have been initiated. The display of post-execution information (e.g., improvement depth, position coordinates, motor current values, and cement slurry flow rates) using a 3D model by enabling the visual confirmation of how the work is implemented is possible. In addition to execution records, the conduct of geophysical and other investigations after execution to obtain actual location information is important. These technologies are expected to be increasingly developed in the future.

In addition to the use of information technology in execution, its use in the investigation and design phases can be conceived. For example, in the investigation phase, big data may be constructed from the results of past studies, and information can be obtained through relatively inexpensive investigations. Then, artificial intelligence (AI) may be used to provide advice on how and where to implement actual investigations. The correlation data of the physical properties of the ground before and after soil improvement can also yield important information. Skilled engineers judge the method and location of investigations based on knowledge and experience, and information technology enables them to accomplish the foregoing with improved objectivity and certainty. In addition, the values used for design are determined from the investigation results. Hence, by allowing AI to learn the relationship between the results of in situ and elemental tests and design values, designers can make decisions based on information to a certain degree. However, as mentioned in the section on execution, overly trusting in information technology is not advisable; this is true even in the investigation and design stages. Current AI only selects the most plausible options based on past data; it does not make technical decisions regarding the construction of civil engineering structures. Ultimately, engineers must understand the situation and make decisions based on knowledge, experience, and theories. Hence, reliable and economical ground improvement can be implemented using information technology as a tool that enhances the skills of engineers.

6.4 ENVIRONMENTAL IMPACT REDUCTION

Reducing the environmental impact of ground improvement will continue to be a challenge in the future. Presently, priority has been given to economic development; however, the pursuit of a dual goal, economic development and environmental protection, is necessary. The United Nations has presented sustainable development goals. Ground improvement contributes to economic development; however, thus far, design has not been well implemented from the perspective of environmental impact. In the future, the sectors involved in ground improvement must contribute in this regard; once such contribution is the achievement of carbon neutrality. Carbon neutrality is a concept whereby the amounts of carbon dioxide emitted and absorbed are the same when any product is created or when a series of anthropogenic activities is conducted. Achieving carbon neutrality is a global concern, and the problems associated with ground improvement must be resolved. During ground improvement, CO_2 is emitted not only by the execution equipment but also by the transport and manufacture of improvement materials; these emissions must be reduced comprehensively. For example, electric power using clean energy can be used to drive the execution machines, the amount of improvement material used can be reduced, the volume of improvement can be decreased, and the execution period can be shortened. Reducing the amount of cement used can be effective, and replacing some of the cement with slag is worth considering. As an alternative, cement with low CO_2 emissions during production can be developed. Moreover, strategies for absorbing CO_2 have been already devised in the cement industry; the same technology can be applied to ground improvement. Thus, environmental impact reduction is another challenge that must be resolved in the implementation of ground improvement.

Reference

1. Takahashi, H., Omori, S., Asada, H., Fukawa, H., Gotoh, Y., and Morikawa, Y. 2021. Mechanical properties of cement-treated soil mixed with cellulose nanofibre, *Applied Sciences* 11(14): 20.

Index

Printed in the United States
by Baker & Taylor Publisher Services